Hair

THE LONG AND THE SHORT OF IT

Art Neufeld

Copyright © 2012 Arthur H. Neufeld, Ph.D.
All rights reserved.

ISBN: 1467953571
ISBN 13: 9781467953573

*For my family and friends,
with thanks for always being supportive,
no matter what I wanted to do.*

*"I said: 'Stop dyeing your hair!'
Now you've no hair left to color."*

*-from a poem in Latin
by Ovid to his mistress, Corinna,
written more than 2,000 years ago*

Table of Contents

Introduction	ix
Chapter 1. FROM FURRY HEADS TO HAIRY HEADS	1
Chapter 2. THE HAIR ITSELF	13
Chapter 3. HAIR GROWTH AND THE CONSTRUCTION ZONE	21
Chapter 4. TYPES OF HAIR: FROM TIGHTLY COILED TO STRAIGHT	33
Chapter 5. NATURAL HAIR COLOR -OR- WHY ARE THERE REDHEADS?	53
Chapter 6. THE GRAYING OF THE HEADS OF THE WORLD	67
Chapter 7. BALDING (HIS AND HERS)	81
Chapter 8. FACIAL HAIR (HIS AND HERS)	105
Chapter 9. HAIR CARE, TRICHOLOGY, FUNGI AND NITS	125
Chapter 10. CUTTING HAIR AND BLOODLETTING	141
Chapter 11. A MORNING AT THE HAIR SALON	155
My Conclusions	173
Afterthoughts	175

Appendices	177
1. The truth about hair myths	179
2. Where hair is not wanted	183
3. Hair expressions and phrases	185
Acknowledgements	187
References and FYI	189

Introduction

Whether you are a man or a woman, you think about it every day. You look at it from the front, from the sides and from the back. You wet it, shampoo it, rinse it, dry it, comb it and brush it. When you look in the mirror, you like it, usually. You describe it as one of your best features and you will not leave home until it is perfect. Still, there is often something about it that could be a little different.

You make regular appointments, noted on your calendar, to maintain it. Cutting it, shaping it, styling it, perhaps coloring it must be just so. Money is not an object. If your favorite barber or stylist moves on, you are lost. Or, maybe you do it yourself. But, you do shop for it and look for the best care products.

You play with it. You push it this way and that. If it is long, you can hook it over your ears. You may toss it back when you are talking, if you have enough. Sometimes you contemplate it or search for gray ones. If you are going to a party, you may twist it, spike it, dishevel it or slick it down, or up. Add color highlights, sparkle or a chic adornment.

It can hint at your age, but you can cover that up. Some days it will infuriate you because it just won't behave. Sleep in the wrong position and you won't even recognize it. Other days, with the wind in it, it will remind you of how free you are.

You agonize over it. You fear losing it. You worry that it will affect your attractiveness. Of course, you always have a comb or brush handy, and check it before leaving the restroom of a crowded restaurant.

You can pin it, tie it up with ribbons, bind it with rubber bands, rest your glasses on it, stick pencils in it, pull it, wrap it, part it, bob it, braid it, frizz it, blow-dry it or ignore it. Whatever you do, it will always be with you.

Or, you can get rid of it, temporarily. Shave it all off, or shave off parts of it and leave other parts. It will make a statement for you.

What is it – it is hair, of course.

But, what do you really know about it?

1

FROM FURRY HEADS TO HAIRY HEADS

"Forget not that ... the winds long to play with your hair."

-Khalil Gibran

Have you ever seen a chimpanzee getting a haircut? For that matter, have you ever seen any furry mammals - apes, monkeys, dogs, cats, horses, cows, gazelles or zebras - in need of tonsorial grooming? Of course not. That's because hair needs to be cut, but fur does not need to be cut. ***Hair is different than fur.***

Only humans have hair on their heads that keeps on growing, giving us the luxury of wearing our hair at different lengths, colors and hairstyles. The hair in our scalp grows and grows throughout our lifetime. Having continuously growing hair on our head is a uniquely human characteristic.

Yes, there are certain competitive canine contestants whose fur may be trimmed or cut as they vie for winning Best in Show. And, yes, certainly, sheep get shorn. But, if the fur of these animals and any other mammals was not cut by their owner, the fur would grow to a certain, short, pre-determined length and then stop growing. The lengths of the hairs of the fur would be relatively even and in proportion to the fur all over the body in a very pleasing and natural manner. Not human head hair.

Human head hair will grow until it is very, very long. And, it is not in proportion to hair on the rest of the body. There is an overall furry appearance found on chimps, dogs, cats and other mammals. Humans are not furry. Imagine a man, naked, with modest amounts of body hair on his legs, arms and chest, and some patches of pubic hair and underarm hair. Now imagine that man if he had neither cut his hair nor shaved his beard in years. A warm, cuddly mammal with a naturally proportionate, attractive coat of fur? No way.

A VERY CLOSE RELATIONSHIP

In our late teens or early adulthood, we look at ourselves in the mirror and our brain starts processing what it sees into an imagined self-image. Maybe, eyes could be bigger, nose could be smaller, mouth could be stronger and ears could be flatter. But, everyone starts doing their hair and the emotional attachment to the fibers growing out of our head has begun. The feelings that develop can be positive or negative. You like your hair or you don't like your hair. Without really thinking that the qualities of your hair are products of your genetic heritage, you take on what to do about what you see. Depending on the hair that you are starting with, you will be more or less successful, and more or less frustrated.

Our genes govern the body in which we exist. Nevertheless, on our head, we enhance or overcome the biology that our heritage has given us. Think about how limited we are to change different parts of our body. What other body part, besides the hair on our head, can we do so much to? Cosmetic surgery can remodel facial features, tummy or thighs, but that's nothing compared to what we can do to our hair. We battle the work of those genes that have produced our hair.

The hair on our head is always on our mind. Hair is an explicit expression of our psyche, which we present for all to see. We beautify our hair to increase our self-esteem. Yet, preoccupation about our hair is a constant annoyance, conjured by our self-conscious mind, and caring for hair is a prime outlet for our self-indulgence. Hair is the crowning signal of who we are and a window to our inner self.

Head hair gives humans an opportunity to demonstrate that they have free will. We invented shampoos, conditioners, curlers, hair dryers, scissors, razors, shavers and trimmers, indeed the entire hair care industry, to exercise and to satisfy our free will. Man or woman, you can grow your hair long or cut it short, color it as you wish and style your hair in the latest look, or not. Cutting, coloring and styling your hair are choices. If you are a man, your facial hair can be grown long to adorn your face as a moustache and/or beard. Shaving or trimming your beard are options. But, all facial hairs are not equal. If you are a woman, facial hair is not considered an adornment, leaving you no choice but to annihilate it.

We are very attached to our hair continuously growing hair. We are drawn to those filaments and it becomes difficult to stop fussing with them. When you pass a mirror, you check your hair first. At the right length, color and style, those strands can be taken from uniquely human to uniquely you. What a glorious, up-top endowment nature has given us. What an opportunity to show the world what we have. Done just right, on the evening of a big occasion, someone important might notice.

HAIR VERSUS FUR

Look in the mirror. See those fibers coming out of your head that you spend so much time on and that must be just right before you leave the house each morning? Now, look at your arms. That thin, often barely visible fuzz is not anything that you give much thought to and certainly not something that you groom with pride. Humans do not attend to the hair on their arms and legs with the time, money and resources used for the hair on their head.

In general usage, the term "hair" refers to both the thick, long filaments on the human head and the fine, thin filaments that grow from the skin in many other places on the human body. However, what we humans call body hair never needs to be cut to maintain a certain length.

Human body hair (underarms, chest, back, arms, legs, pubic) grows to a predetermined short length and then falls out. Human hair on the head is hair; hair on the human body is actually more like fur.

We use the term "fur" for the fine, thin growths from the skin that can be found covering our favorite pets and most mammals. Certainly, the hairs in the magnificent mane of the male lion, or in the silky mane of a horse, grow long. But still, they grow to a certain predetermined length. These animals with exceptionally long hairs have developed a significant display with their body fur, but it is not really hair like humans have hair.

In furry mammals, all of the hair follicles work together and many mammals lose some of their fur seasonally. When the weather gets hot, animals in the wild (and in your house) shed fur. With winter coming on, the coats of bears, squirrels and other northern mammals become thicker. The hair follicles in these coats of fur are in synchrony. The hair follicles in humans, in the scalp and the face, are not in synchrony. Each human hair follicle is independent, produces a hair randomly and does not participate in a seasonal rhythm. Humans do not lose their hair seasonally nor does their hair get thicker seasonally. These are even more differences between fur and human hair.

Is human pubic hair, hair or fur? Clearly, by our definition of hair as continuously growing, pubic hair is fur. People do not have to cut their pubic hair every 4-6 weeks. Of course, there are those who shave off their pubic hairs, but this is in pursuit of eroticism not just trimming the bush. Pubic hair is a truly elaborated, highly differentiated in appearance, form of fur that highlights the genital organs, and captures and releases the intoxicating scent of human sexuality. Drawing attention to the genital organs is common in many species and would be in humans if we ran around naked.

Application of the term "hair" needs to be reassigned. We humans have "hair" on our heads and, in reality, "fur" on the remainder of our bodies. Animals have fur all over their bodies. When we snuggle up with our pet dogs, cats, guinea pigs, gerbils, mice or fellow humans, we are snuggling fur to fur. The standard definitions for "hair" and "fur" as related to the human body really should be changed.

Whether a modification in usage of the term "hair" can be incorporated into thought, speech and writing remains to be seen. Referring to hair on the head is understood, but will men ever refer to growing fur on their chest or will women ever refer to shaving the fur on their legs? Probably not. Throughout the remainder of this book, "hair" refers to the filaments on the head (scalp and facial). The so-called hair on the surface of the human body will be referred to as "body hair," but please think about it as fur. Human body hair is a remnant of our furry ancestors.

EVOLUTIONARY LOSS OF FUR

The evolutionary loss of a significant furry coat on the human body, as compared to other mammals, has been a subject of longstanding interest to anthropologists, people who study how we became human. Monkeys, apes and chimpanzees have dense fur all over their bodies; we do not. The loss of thick fur covering the human body occurred sometime after the evolutionary split, 4-8 million years ago, of our common ancestor into two primate lines that led to chimpanzees and to people. Genetic evidence suggests we lost our body hair about one million years ago. Many explanations for the loss of fur in humans have been put forth.

Most likely we lost our fur because of the need for better thermoregulation. As we evolved from inhabitants of the jungle to hunters in the sun-soaked savannah, keeping the burden of full body fur may not have been advantageous for maintaining body temperature at 98.6° F. Maybe we were too hot. Interestingly, the body temperature of many furry mammals is higher than human body temperature. A higher body temperature may not have been optimal for peak performance of our brain, heart and muscles. Losing fur probably brought human body temperature down to produce a better, more energy efficient lifestyle. Furthermore, anthropologists have associated the loss of fur with the development of sweat glands in our species, another uniquely human characteristic. Sweating with minimal body fur apparently provided a better means for thermoregulation when we were running after big game.

But better thermoregulation does not explain the evolutionary development of long human head hair. We lose much of our body heat via the top of our head. Wearing a hat keeps the head and body warmer. Long head hair will make the head and, consequently, the body feel hotter. That might be a great plan for people who live in northern climates, but people who live in warmer climates developed the same long thick hair. Long hair is a burden when temperature rises. When men or women with long hair exercise or are in hot climates, the ponytail appears to keep hair off their neck.

There have been suggestions from a few anthropologists that as humans evolved we went through an aquatic proto-hominid lifestyle. Perhaps we were once dolphins with legs and arms. Dolphins and whales are mammals that do not have fur but have thick layers of fat that help keep body temperature constant. If we evolved through a water stage, presumably we would have lost our body fur and become fatter. However, there is not much support in the fossil record for early humans swimming around as denizens of the deep. In any event, if humans did spend an evolutionary stage in the water, head hair would have reduced swimming speed. Olympic competitive swimmers shave their heads to improve their speed. Swimming around with long hair might look nice on mermaids but it would definitely not have been good for our evolution.

Charles Darwin and his followers have always had a problem with fur. There is no scientific evidence that the specialized hair/hair follicle structures that produce the furry coat of mammals evolved from something else on the skin or from a transitional structure. The something else could have been like scales on the skin of reptiles. The transitional structure could have been, perhaps, a whisker-like structure that had some other function. But there are no precursor skin structures in the fossil record for hair follicles and the hair they make.

For a Darwinist, fur came out of nowhere. "Out of nowhere" is not good for the evolutionary theory. Darwinists like to trace evolutionary development. The brain of a lobster is relatively simple, the brain of a reptile has some sophistication and the brain of a human is much, much more complex. Darwinists have found fossil remains of fish with fins that have become limb-like, suggesting that fish could have crawled out of the water and evolved into land animals with legs. The absence of any demonstration of the evolution of hair in the fossil record is fuel for fundamentalists who argue against evolution. Apparently, hair just appeared on the skin of mammals without any previous evolutionary trace. One point for the Creationists; zero for the Evolutionists.

Darwin did promote the hypothesis that the loss of body fur in humans was favored by potential mates and, therefore, used in sexual selection. As we evolved, less body hair was seen as more attractive. But why would less body hair be more attractive to an early hominid? Certainly, male and female chimpanzees find each other attractive and they seem to be having sex all the time. Obviously, a mammal does not have to lose body fur to be sexually attractive. But we did. As humans lost body fur, there was a concomitant specialization and growth of pubic hair to bring attention to our sexual apparatus. And then we started wearing clothes.

The great Darwin did not take on trying to explain the evolution of human head hair. But, we can try to extrapolate Darwin's thoughts on sexual selection to the hair on our head. Human head hair can be made attractive and much of our personal resources are spent on making it so. However, both men and women can have long or short hair, so head hair provides inaccurate information on gender. Also, the shape, color, length or style of hair on the head gives no definitive biological signal. In birds, plumage is associated with specific, desired qualities of a mate; no such association is known for human head hair. Thus, extrapolating Darwin's ideas are not particularly helpful.

British scientists have presented an intriguing recent hypothesis that relates the loss of body hair in humans to the reduction of our parasite load. All of those annoying little critters, like fleas and lice, which find a home in the fur of animals are greatly reduced on the body of a relatively furless human. This rationale is compelling for human body hair, but does not help with the explanation of human head hair. Long, ungroomed human head hair can be a hospitable haven for many parasites.

None of the hypotheses about loss of body hair in humans deal with human head hair and human body hair separately. These hypotheses only relate to human body hair as fur. Actually, the continuous growth of human head hair is not consistent with any of the hypotheses on why we lost fur.

WHY DOES HUMAN HEAD HAIR CONTINUE TO GROW?

No one has pointed to any anatomical differences between the hair follicles on the scalp and the hair follicles on the body that can be linked to continuous growth of hair. On the other hand, it seems as though no one has looked carefully. Textbooks on dermatology, the study of the skin, do not present separate information on the hair follicles of the head and the hair follicles of the rest of the body. In fact, when reading textbooks on human dermatology, for example the relevant chapters in *Rook's Textbook of Dermatology*, it is not possible to know whether the diagrams and descriptions in the text are based on studies of the skin of the head or the skin of the limbs. Perhaps there are anatomical distinctions comparing the hair follicle in the scalp, the hair follicle in the beard and the hair follicle on the arm. In the medical community, dermatologists are supposed to be the most knowledgeable about hair. However, there is little training or experience regarding hair in this specialty. Thus, there is not much anatomical or clinical knowledge when it comes to distinguishing the all-important hair on our heads from what is left on our body from our furry past.

Scientists who do experimental research often use laboratory animals to model the physiological system of interest and study how that system is regulated. Sometimes the results of these mouse experiments are announced with great fanfare. You may see a newspaper headline touting the latest finding on hair growth, or hair loss, in laboratory animals as promising a cure for baldness. But extrapolating laboratory findings on furry animals for relevance and application to human hair remains to be accomplished. Because animals have dense fur on their heads, not continuously growing hair, we cannot study head hair growth in a laboratory mouse or any other furry mammal. There are no good, reliable animal models to study the growth of hair on the head. However, research continues. Based on past and future laboratory animal and test-tube experiments, we can postulate what may be happening in the human scalp, but the true demonstrations of the reasons for, and the mechanisms causing, hair on the human head to grow, or not grow, await investigations by clever scientists.

We do know that the hair follicles on the head are high performance factories. Production of the hair shaft in the hair follicle occurs during a specific phase of activity, anagen. To produce human body hair of a short length, the anagen phase is regulated to operate over a predetermined period of a few months, during which time the body hair grows to a certain length and then stops growing. Human head hair has no such limit. The duration of anagen for human scalp hair is not temporally regulated, meaning that there is no timing device that tells when hair growth should start and when it should stop. Remarkably, anagen can last 6 years or more for hair on the head. Six years! Clearly, human head hair follicles have a greatly extended anagen phase as they work day after day without a rest year after year. The hair follicles in your head are another marvel of the human body. Nevertheless, functionally, why human head hair has broken through the temporal regulation that carefully controls the length of human body hair and animal fur, we do not know. More on the phases of the growth cycle of hair follicles in Chapter 3.

HAIR TRANSPLANTS PROVE THE POINT
Proof-of-concept can come from the oddest places. As stated above, we cannot directly study the growth of head hair in laboratory animals. Ideally, experiments should be carried out in humans to establish that there is something special about hair on the human head that makes it grow differently than hair on the leg. Experimentally, a scientist would approach this kind of question by performing transplants: take a patch of

skin from the arm and transplant it to the head. Fortunately, experiments in humans are not necessary; hair is transplanted every day.

The success of human hair transplantation demonstrates that, indeed, the growth of human head hair is different than the growth of human body hair. The length of hair that is produced by a transplanted hair follicle is dependent on the location on the body from which the hair follicle was taken. The field of modern hair transplantation is based on the concept of donor dominance. This concept indicates that for hair follicles to grow long head hair after transplantation, the hair follicles must be taken from regions of the body that normally grow long head hair and not from regions of the body that grow body hair.

In other words, the neat implants on the hairlines of balding men are populated by transplantation of hair follicles from neighboring, non-balding regions of the head. Hair follicles from the chest or legs that are transplanted to the head will continue to grow the type of short body hair typical for the chest or legs. Transplantation of chest or leg hair follicles will not result in restoring long hair to the head that can be combed, brushed or styled. Conversely, scalp hair, transplanted to the leg grows considerably longer than does the indigenous surrounding leg hair. The growth of human head hair is intrinsically different than the growth of hair on the rest of the body.

THOSE WHO CAME BEFORE
The uniquely human characteristic of long, continuously growing head hair must have developed before our particular ancestral tribe migrated out of Africa 50,000-70,000 years ago and began to populate the world as the human species. Therefore, it is likely that many earlier hominid lines had continuously growing head hair.

About two million years ago, the bodies of our early hominid ancestors started to become more human-like. We were no longer chimps. The human ancestral line undoubtedly had continuously growing head hair by the time of the population split between Neanderthals and modern human ancestors, about 250,000-500,000 years ago. According to the recently deciphered Neanderthal genome, when these hominid relatives were hunting and gathering in Europe, they probably had hair like we do and some Neanderthals actually had red hair. This dating means that between one million and five hundred thousand years ago in Africa, hair that grows on the head became qualitatively different than what was left of the fur on the other parts of the human body. We don't really know exactly when in this time frame hair started to grow, but tending to the hair on the head has become an obligatory human chore ever since.

This brings up an interesting question. Why was the hairdo invented? Undoubtedly, head hair grew first and then one fire lit night an amorous cave-mate said: "Why don't you do something with that stuff on your head?" After that momentous evening, hair styling was in. Nevertheless, we are still at a loss for why long head hair grew there in the first place.

AND TODAY

Whatever hair is for, we treat it terribly in the name of making it look good. The stresses, physical processes and caustic chemicals that we use on our hair would never be used on any other part of our body. You would not wash with detergent, apply lye or dry with scorching heat any part of your anatomy. Maybe it is good that hair is dead.

Washing hair with shampoos removes the natural protective oils and leaves the hair long, limp, heavy and absorbent. Conditioners make the hair greasy. Coloring uses toxic dyes that coat the hair or penetrate into the hair and would probably cause cancer if ingested. Or, coloring can be achieved on a more permanent basis by applying chemicals that degrade the pigments that were there originally. Straightening hair relies on chemicals that are strong enough to denature the very fabric of which hair is made. Extreme heat achieves similar results by cooking hair into the preferred style or look.

These corrosive chemicals and these levels of heat would burn holes in your gut. None of these treatments could be tolerated anywhere else in your body. We are traumatically injuring one of our most alluring features in the quest for attractiveness. We do it willingly and we pay for it.

Go to the mirror again and contemplate your hair. There is so much more to know about something that is so familiar. What is hair? How is hair made? What determines the amount of time that hair grows? What determines the length to which hair grows? Why are there different types and colors of hair? Why do we lose hair? Why does hair turn gray? What are we doing to our hair when we groom and beautify this highly visible characteristic? This book is all about the hair on your head. Grab a single hair, shrink yourself down to become a tiny explorer and slide down the hair shaft to the scalp. Look around. There is much to learn. First, let's examine the structure of a hair and what it is made of.

2

THE HAIR ITSELF

"The hair is the richest ornament of women."

-Martin Luther

The term "hair" can refer to a single hair shaft, above the skin surface, or it can be pluralized when referring to hairs left on the sides of a bald man's head. However, most often, "hair" is understood to be pluralized and refers to all of the hairs on the head. The word "hair" came to the English language from the Anglo-Saxons *haer* (which came from a Dutch influence: *haar*) and first appeared around 1000 CE in an Old English text referring to *haer* on the *heafde*.

THE NUMBERS IN YOUR HAIR

The rate of hair growth on the head varies, slightly, amongst individuals, between ethnic groups and by location on the scalp. On the top of the head, hair grows about one fiftieth (1/50th) of an inch in 24 hours. At the temples hair grows a little slower than on the top of the head. Generally, hair grows slightly faster in women than in men and some people's hair grows faster than others, but there are wide, overlapping ranges for the rate of hair growth. Calculating out the average, head hair grows about 1/2 inch per month or about 6 inches per year. Consider this: For shoulder length hair, which is about 12 inches long, the hairs have been growing for two years. The split ends at the tips of shoulder length hair were emerging from the scalp as intact hairs two years ago.

The longest documented, and presumably still growing, hair in the world is 19 feet long, according to the last time anyone made an entry in the *Guinness Book of World Records*. This record belongs to Xie Qiuping of China, who started growing her hair long in 1973 at the age of 13. When the former record holder, Tan Van Hay of Vietnam, died at age 79, he had hair that was 23 feet long. For many years Tan did not wash his hair. According to his wife, he let his hair grow for more than 50 years because he often became sick after a haircut.

If you live to be 75 years old, you will have grown, from a single hair follicle, approximately 450 inches (38 feet or 13 yards) of hair during your lifetime. That may not seem like much but let's continue to calculate. You have 100,000-150,000 hair follicles that gradually decrease in number with age. Starting with, for example, 110,000 working hair follicles and allowing for age-related loss but excluding bald heads, by age 75 you will have produced approximately 50 million inches of hair from all the hair follicles in your scalp. That is over 4 million feet or 1.4 million yards of hair. Does that sound like much? It is the equivalent of over 750 miles. The total amount of hair that you produce in your 75 years, laying each hair end-to-end, would reach from New York City to Chicago.

If you cut a piece of hair the way that you would cut a salami and look at the cross section, the diameter of a single hair shaft is, on average, about one three-hundredth (1/300th) of an inch. The term diameter is not really applicable because in cross section, hairs are not always truly round. As explained in Chapter 4, people from different geographic regions have different hair types and shapes. Therefore, the diameter or width of a human hair does not have a standard value and can have a range of tenfold, anywhere from less than 1/1000th of an inch to about 1/100th of an inch in cross section. Genetics have a lot to do with hair diameter, as well as age. When you were a baby, your hair was thin and fine. When you grew up, the width of your hair thickened. As you get older, the cross sectional area of your hair will thin again. Someone in the 17th century may have measured the width of a hair and then gave some room to spare to the meaning of "hair's breadth." At that time, a hair's breadth had an actual value and was defined as 1/48th of an inch, which is about six times thicker than an average hair. Still, a hair's breadth is indeed a very narrow margin. Othello (Shakespeare) speaks of his near death escapes: "Of hair-breadth 'scapes i th' imminent deadly breach."

Each hair has a surface area. To picture surface area, think about using a thin, sharp knife to open a hair, straight along the length of the hair shaft, and laying the total surface out flat as a sheet. The total surface area of a full head of 110,000 hairs that is shoulder length is about 10 square yards. Ten square yards of fabric would reupholster a very large sofa. That is a lot of surface area to cover during daily shampooing.

A single strand of hair, 6 inches in length, weighs a little less than 1 milligram, which is approximately one thirty thousandth (1/30,000th) of an ounce. Assuming 110,000 hair follicles on the head, excluding those who go bald, but allowing for age related loss, by age 75 the average person will have produced about 15 pounds of hair in their lifetime.

Most women are walking around with less than a half-pound of hair on their heads; most men have far less weight of hair on their heads. The reason: haircuts. We can estimate that over your lifetime, to maintain your hair a certain length you have left about 15 pounds of hair on the floor of your hair salon or barbershop. The fractional amount on the floor, as you have your hair cut every month or two, can easily be disposed of by your haircutter. However, on the floors of barbershops and hair salons, and homes where people cut their own hair the world over, the numbers get quite large. In any given year for the US population, there is as much as 60 million pounds of hair on the floor. In cutting the hair of the world's population, there is as much as one billion pounds of hair on the floor each year. In the US, most of the cut hair gets swept into a black plastic bag, put in the trash container and ends up in the garbage dump of the local municipality. Fortunately, garbage dumps are active places. Earthworms eat the discarded, protein-rich hair.

Is there a use for this wasted, natural resource? Actually, there is. Cut human hair is absorbent and has been used to sop up spilled oil. During the disastrous British Petroleum oil spill in 2010 in the Gulf of Mexico, a call went out for hair. There was a huge response when hair salons put up signs saying: "Help the Oil Spill. Get a Haircut." Around the US, barbershops and hair salons were sweeping up the hair on the floor and sending it to a California organization of environmentalists. Perhaps a million pounds of hair was collected. The purpose was to make the hair into booms (hair stuffed into long nylon tubes) that would be used to absorb the oil. Such booms have been used in coastal areas and shallow water to protect marshes and wildlife. Unfortunately, after a period of active hair collection, BP announced that hair filled booms would not be useful in this deep water spill.

THE HAIR (OR HAIR SHAFT)

Technically, the filament that you see emerging from the scalp is called the hair shaft. For simplicity, we call it the hair. There is the part of the hair above the skin surface and a part that penetrates into the upper layer of skin, the epidermis. On average, there is a density of about 1600 hairs per square inch of scalp. However, the density is not uniform and can vary on different regions of the head. Your hair may feel thicker in some places compared to other places on your head. Hairs can emerge from the scalp very close together, appearing like two hairs coming from a single hair follicle, but actually growing from two fused hair follicles.

The fine structure of a hair can only be seen with a microscope. Using magnification, we see that everything is dead. Because all of the parts of a hair are dead, the hair shaft is not metabolically active and, therefore, does not need blood vessels supplying nutrients. Also, because the hair shaft is not alive, it does not respond to stimuli.

In cross section, a hair is divided into three concentric zones. Each zone is continuous and runs from the root of the hair to the tip of the hair. Starting from the outside of the hair shaft:

- The *cuticle* is the tightly formed, multilayered, outer structure of the hair shaft made up of flat, thin overlapping, ghost-like cells. The cells of the cuticle are closely attached and are laid out like roof shingles or fish scales. The cuticle is made up of 7-10 concentric layers of these shingles, one layer on top of another layer. The overlapping edges of the cuticle cells in each layer are oriented outwards, away from the head and towards the tip of the hair shaft, thus facing away from the root.

  The cuticle has a smooth appearance and is important to how hair looks. The layers of the cuticle reflect light, giving the hair sheen, and minimize friction between hairs, giving the hair manageability. Also, an intact cuticle protects the hair shaft from chemical, physical and environmental insults. If the cuticle is damaged by physical breakage, chemical removal or oxidation, the hair becomes more porous and vulnerable. Often, damaging the cuticle is the first step in hair coloring and styling.

- The *cortex* makes up most of the thickness of the hair and gives the hair its length, its width and its mechanical properties. In the cortex are closely packed remnants of dead keratinocytes and lots of keratins and pigments surrounded by a matrix. Keratins give hair strength and pigments give hair color.

- The *medulla* is the disorganized, open area at the center of the hair shaft. In the medulla, everything is very loosely connected and partially separated by air spaces. The medulla contains pigments and the protein, trichohyalin, but little or no keratin. Slight differences in the protein trichohyalin, caused by variants in the gene, produce wavy or curly hair. Those natural curls that you may want to straighten are given their shape by the very center of the hair shaft.

ARCHITECTURAL SUBSTRUCTURE

Keratins are the structural proteins that are assembled into the building blocks for forming the hair shaft. These relatively small proteins have a corkscrew, helical shape. In hair, there are 16 slightly different keratins formed by the combination of chains of different amino acids. Four helical keratin proteins are twisted together into units called protofibrils. In the terminology of physics, protofibrils are referred to as supercoiled coils.

Protofibrils are the basic building blocks that make up a strand of hair. Each protofibril building block is tiny and cannot be seen with even the most powerful microscopes. Nevertheless, when millions of protofibrils are assembled together, a visible, touchable, twirlable strand of hair is made.

One of the amino acids in keratin, the sulfur containing amino acid cysteine, plays the key role in holding together the keratin molecules to form the protofibrils and also in the cohesion of these building blocks to form the hair shaft. Imagine each keratin protein as a corkscrew with protruding muscular arms sticking out. Those arms with rippling muscles are tightly clasping hands with other muscular arms sticking out of other keratin corkscrews. The two hands form a cysteine disulfide chemical bridge. The grip is firm and cannot be easily broken.

Many two-handed grips between foursomes of keratins form a protofibril. Protofibrils are attached to other protofibrils also by the protruding muscular arms as cysteine disulfide chemical bridges. These chemical bridges between the protofibril building blocks are repeated over and over again, throughout the length and width of the hair shaft, and bind the protofibrils together into microfilaments, which are like short, thin fibers. Then, like making a rope or yarn, hundreds of thousands of microfilaments are woven together into macrofilaments, which are larger, thicker and longer. The macrofilaments are bundled by the tens of thousands to create a single strand of hair. At all levels of organization and assemblage, the keratin molecules, the protofibrils, the microfilaments and the macrofilaments are in parallel, and oriented in the same direction, from root to tip, to produce the long hair shaft, to give the hair shaft thickness, and to endow the hair shaft with strength, flexibility and durability.

To examine how lifeless filaments hanging from your scalp can be made and can grow long, let's explore the construction site where a hair is assembled. Put on your hard hat and we will descend deep into the hair follicle.

3

HAIR GROWTH AND THE CONSTRUCTION ZONE

*"Rapunzel, Rapunzel let down your hair,
so that I may climb the golden stair."*

-Brothers Grimm fairytale

The hair shaft, which has no living parts and is made up of inert, complex proteins, is constructed by specialized living cells in the skin, that are some of the most active, hard working and tireless cells in the body. They have their own independent work schedules. For whatever reason we have hair, we grow a lot of it. In this chapter, we will look into how hair gets to be hair.

THE HAIR FOLLICLE
The hair follicle is the factory that constructs the hair shafts growing from the skin. Actually, hair follicles are distributed all over the human body, except the palms of the hands and the souls of the feet. We are born with five million hair follicles and that is the largest number of hair follicles a human will ever have. We do not make new hair follicles or increase the density of hair follicles anywhere on our bodies or at any time during the course of our lives.

Populating the skin of the scalp are about 110,000 hair follicles in a newborn infant, which are densely packed, fairly close together. The density of these hair follicles decreases as we grow from being a baby to old age. The initial reduction in density is because our head expands as we grow up, but new hair follicles are not added. Imagine blowing up a balloon halfway and then, with a black felt tip pen, putting black dots spread evenly over the balloon's surface. When you blow up the balloon to its full size, the black dots will be further apart from each other. In the fully blown up balloon, the black dots are at a decreased density. This is similar to going from babyhood to adulthood as the size of the head increases. However, from adulthood into old age, there is further reduction in the density of hair follicles. The head is no longer expanding, but hair follicles are dying off. Decreased density means fewer hair follicles per square inch. Consequently, hair on the fully formed head thins with age.

The hair follicle is an organ. Other organs include your brain, heart, liver, eyes and kidneys. The hair follicle organ has a blood supply, an internal lining of cells, an innervated muscle, a sebaceous gland and a cup of stem cells that sometimes are resting, sometimes are very active and sometimes are dead. Each hair follicle organ has a strenuous life cycle; it works hard for years to build the hair shaft, then shrinks, becomes inactive and rests, and later resurrects itself into a fully functional organ. The brain, heart, eye and kidney cannot regrow themselves. No other organ in the body has the ability to completely rebuild itself. The hair follicle is a unique organ that can renew itself over and over again.

Each one of the remarkable hair follicles in your scalp is a separate, independent, autonomous organ. Think about a single blade of grass in a lawn. That blade of grass is growing as happy as it can be because as far as it knows, it is the only plant in the world. It is not in communication with the individual grass plants next to it and there is no external, overall control for the growth of each blade of grass in the lawn. Every blade of grass does its own thing and a beautiful green lawn is produced. Just like a plant producing a blade of grass is autonomous, each hair follicle has its own cycle and schedule, and doesn't take orders from anyone.

The hair follicle is located beneath the skin surface as a cylindrical, vase-like structure of the epidermis that extends down into the dermis. At the base of the hair follicle is the construction zone where capillaries nourish the working cells, the keratinocytes and the melanocytes. This entire structure, at the base of the hair follicle, is called the bulb.

The hair follicle undergoes cyclical changes in size, activity level and structure. A fully active hair follicle, building a hair shaft, is 3 times larger, mostly longer, than a resting hair follicle. When active, the cells of the bulb divide every 1-3 days, faster cell division than most other cells in the body. The activity in the construction zone can go on for years before subsiding into a relatively short period of dormancy. When awakening from its dormant state, the cells of the hair follicle release, locally within each hair follicle, many chemical signals that orchestrate the self-rebuilding of the hair follicle into an active organ. Rebuilding and reactivation of the construction zone of the hair follicle is not due to extrinsic factors from the bloodstream or the environment. Control is local, intrinsic and, again, autonomous.

When a hair shaft emerges from the hair follicle, the hair grows in a specific direction. Think about it. People have hair that grows away from the face, front to back on the top of the head, and down the sides and back of the head. These directions are due to the slant, or angle, by which the hair follicle is situated and pointing in the skin. Imagine what a head of hair would look like if there was no directionality to each growing hair shaft.

Hair also whorls. A hair whorl is a patch of hair growing in the opposite direction than the rest of the hair. A cowlick, such a graphic word, is a whorl. In many people, whorls take a circular pattern on the top of the head and there can be two and even three whorls. Most people who have whorls, whorl clockwise, but about 8% of people whorl counterclockwise. There may be a genetic relationship between counterclockwise whorls and left-handedness, perhaps both characteristics being influenced by a single gene.

There are also studies that relate counterclockwise whorls to homosexuality. A few years ago, a psychologist in California examined the pates of men attending the Long Beach Gay Pride Festival and found support for the hypothesis that the whorls on the crown of gay men swing the other way. Additional studies have confirmed, at least statistically, a higher percentage of counterclockwise whorls in male homosexuals. Women have yet to be studied. In addition, there are several phrenological claims that suggest that hair whorls can be used to determine everything from brain development to temperament to learning ability. In whatever direction hair grows, slants or whorls, the construction of the hair shafts is the same.

THE CONSTRUCTION ZONE AND THE WORKERS

The hair follicle is about one-third of an inch long when actively building a hair with the construction zone for the growth of the hair at the very bottom. This small area contains the active biological materials that construct a hair shaft that can be many feet long. It takes about three weeks for a newly formed hair to appear at the surface of the scalp. After emerging, the hair shaft will keep growing from the base of the hair follicle for years. Here are the construction workers:

Keratinocytes

Keratinocytes are specialized cells in the hair follicle that build the structural scaffold to give the hair shaft length, width and strength. They are the engineers and onsite producers of building material, and multiple generations of keratinocytes in the construction zone continue the work for years. The keratinocytes fill themselves up with structural keratin proteins and internally assemble protofibrils. They gorge themselves on the packages of pigment received from the melanocytes. When their load is ready, keratinocytes elongate and migrate into the nascent part of the hair shaft, eventually becoming spindle shaped as they snuggly place themselves inside the bottom of the lengthening hair. Here they install the microfilaments and macrofilaments to become the superstructure of the growing hair shaft. The keratinocytes give up their lives as a way of melding into the newly formed end of the hair shaft. During the remainder of the existence of each hair shaft, new keratinocytes containing keratin and pigment are only added to the bulbar end of the hair follicle. Keratinocytes do all of their work in the basement as this long structure, easily millions of times taller than they are, keeps growing.

Melanocytes
Melanocytes, the pigmented cells in the body found in hair follicles, skin and eyes, give the hair its natural color. These cells manufacture melanin pigments and work at the bottom of the hair follicle in the construction zone, where they are very busy, contributing melanin to the growing hair shaft. Melanocytes accumulate the melanin in melanosomes, which are like sacks of pigment inside the cell, and extend dendrites, long, branched finger-like tentacles, that contact other cells. The activities of melanocytes are tightly coupled to the hair growth cycle so that they are always working alongside and together with the keratinocytes. To color the hair, melanocytes transfer their pigment in melanosomes into keratinocytes that have placed themselves in the cortical layer of the hair shaft under construction. It is the pigments in the dead keratinocytes in the hair shaft that give hair its natural coloring, from black to blonde (Chapter 5).

Dermal papilla
The dermal papilla is like the lunch wagon at the construction site, but food is available night and day. Located at the base of the hair follicle, the papilla has a rich network of capillaries that are constantly perfusing the hair follicle with blood. The dermal papilla is the source of nourishment (glucose, amino acids, oxygen) for the surrounding, active worker cells in the hair follicle matrix that are building the hair shaft.

Stem cells

Stem cells are capable of dividing and morphing into a variety of highly specific cells. There are, of course, scientific, philosophical and political debates about using stem cells to grow new tissues, organs and even entire bodies. The stem cells in hair follicles are probably not pleuripotent, meaning that they could not easily be used to grow new organs. The role of the stem cells in the hair follicle is to become melanocytes when needed in the resurrection of the hair follicle for the next hair growth cycle.

GROWTH OF HAIR

Hair follicles undergo a repetitive sequence of growth and rest known as the hair growth cycle. At this moment on your head, hair follicles are in one of three stages. Here is the schedule:

Anagen is the active phase of the growth of the hair shaft, which on the human head has a much extended duration. The cells in the hair follicle divide rapidly during anagen as a new hair is formed and lengthens from the base. The hair follicle in the scalp stays in this active phase of growth for many years, growing what is called a "terminal hair." In contrast, human body hair on, for example, arms and legs grows slower and has a very short active growth phase of about 30 days. Normally, 80-90% (about 95,000) of the hair follicles in the scalp are in the anagen phase at any given time. This is the phase of the hair growth cycle that produces the 750 miles of hair over a lifetime.

Catagen is a transitional stage and about 3% (approximately 3,000) of all hair follicles on your head are in this phase right now. Growth of the hair shaft has stopped in these hair follicles. For a single hair follicle, the catagen phase lasts 14-21 days. Catagen is the regression phase of the hair growth cycle in which the upper two-thirds of the long hair follicle, that was active in anagen, withers away and is reabsorbed. There is massive death of cells and clearing of debris from the hair follicle, bringing the bottom of the hair follicle closer to the skin surface. During this phase, what was the terminal hair is now called a "club hair." The

HAIR GROWTH AND THE CONSTRUCTION ZONE

bottom of the hair shaft moves up the shrinking hair follicle towards the surface of the skin.

Telogen is the resting phase of the hair follicle and is usually occurring in 5-15% (about 10,000) of all hair follicles on your head at any given time. This phase lasts for about 100 days for hairs on the scalp. The telogen phase is longer for hairs on the eyebrow, eyelash, arm, and leg; hairs in these locations spend a lot of time not growing. During telogen, the hair follicle on the head is completely at rest and the club hair is completely formed. Eventually, the club hair drops out; the shedding of the club hair is **exogen**. Pulling out a hair in the telogen phase will reveal a solid, hard, dry, white material at the root. Don't fret over a few hairs on your pillow in the morning or in your brush when you brush your hair. It is normal for healthy, non-balding individuals to shed 25-100 hairs from the telogen phase hair follicles on their heads each day. At the end of telogen, after the club hair has been shed, stem cells help regrow the hair follicle and the anagen phase of the cycle starts anew.

Although hair growth during the anagen phase is controlled locally by cells in the hair follicle and is autonomous, there are times when systemic, circulating substances in the blood like hormones can alter hair growth. In males going through puberty, circulating androgens (testosterone and its cousins) stimulate hair growth on the face and body. Androgens stimulate hair follicles in the beard, chest, legs, arms and groin, that have been producing peach fuzz, to grow thicker, darker hair shafts. But note, hair follicles in the **scalp** are not stimulated during puberty, an important point of which you will be reminded when we discuss balding. There is never an increase in the number of hair follicles due to circulating hormones anyplace on the head or the body.

Time to dispel a fairly persistent myth: hairs (and nails) do not grow appreciably, if at all, after death. Because each hair follicle is autonomous, essentially on its own, news of the death of its owner may not have an immediate effect. For a few hair follicles diligently producing hairs in the anagen phase of the hair growth cycle, there may be sufficient, stored nutrients to keep the keratinocytes working for a while. In a short period of time, within 24 hours, the nutrients will be used up and the hair follicle will give up the ghost. Any hair that may grow for 24 hours after death will be less than 1/50th of an inch longer and not really noticeable.

TELLTALE HAIR

Hair has much to tell. Because hair continues to grow throughout your life, and the growing end is close to blood vessels in the dermal papilla, foreign chemicals that are in the bloodstream will diffuse out and get trapped in the base of the hair shaft during construction. Like throwing pebbles onto a drying cement walkway, the substances will be there for a very long time. As the hair grows, the trapped substances slowly move away from the scalp. Thus, the matrix properties of hair and the presence of melanin in hair bestow these filaments with a recent history of your exposure to chemicals that you may, or may not, want known to others. Darker or coarser hair stores chemicals at higher concentrations than lighter hair. If a dark haired person and a light haired person consume the same quantity of a foreign chemical, more of the material will show up in the darker hair.

Hair is a temporal repository for all sorts of chemicals to which you have been exposed. The substances that end up in your hair can be analyzed in specialized laboratories using gas chromatography-mass spectrometry and less than 1/100th of an ounce of hair is needed to perform these analyses. Because hair continues to grow, recent exposures to chemicals since your last haircut can be determined. As said in the trade: hair has a longer detection window than urine, and studies have shown that analyses of hair are more reliable than analyses of urine. Analyzing segments of hair at different lengths from the scalp provides a time course of exposure, or use and non-use.

An extraordinary example of the information that can be derived by examining hair, even old hair, is that of Beethoven's hair. While

HAIR GROWTH AND THE CONSTRUCTION ZONE

Beethoven lay dying in 1827, admirers were snipping locks of his hair as keepsakes of the great composer. He was practically shorn bald when he was buried. At least one of the locks is known to have survived and, in 1974, was auctioned at Sotheby's for $7300. Recently, strands of that hair were given over for chemical analysis and found to contain 100 times the normal, expected amount of lead. Lead poisoning, plumbism, probably accounted for Beethoven's erratic moodiness, depression, constant abdominal pains and may have been the cause of his death. However, it is not thought that plumbism contributed to Beethoven's deafness. Where the lead in Beethoven's hair came from (ingestion, environment) is not known, but foul play is not suspected.

Knowledge of what is in hair is of interest to various law enforcement professionals, environmental toxicologists and forensic investigators. Drug abuse will show up in hair. Cannabis and cocaine are easily detected. Also detectable are: heroin, ecstasy, amphetamine, morphine, codeine, diazepam, benzodiazipine, oxycodone, methadone, tramadol, fentanyl, buprenorphine, propoxyphene and steroids. Use of alcohol can be detected in hair by analyzing it for chemical markers that are only produced when there is alcohol in the bloodstream. Nicotine is detectable in hair. Interestingly, cosmetically altering hair does not interfere with the analyses.

Exposure to substances deleterious to health in air, dust, sediment, soil, water, food and toxins in the environment will also show up in hair. The minerals: mercury, arsenic, lead, calcium, iron, cadmium, copper, chromium, manganese, thallium and zinc can be detected in hair and can be indicative of heavy metal environmental poisoning. Organic and inorganic environmental pollutants are detectable in hair.

Hair can be used for forensic identification, placing an individual at the scene of a crime. Hair from a crime scene is examined for color, curliness, condition of the roots and tips, ethnicity, dried blood, soil and gunshot residue, and then matched with that of a suspect. If the hair sample contains the root and, therefore, cells from the hair follicle, DNA can be determined.

HAIR

There is a lot of forensic evidence in hair. Probably the most famous case is the analytical discovery of very high levels of arsenic in the hair of Napoleon. Apparently, the Emperor was the victim of a conspiracy by the French and English to slowly and deliberately murder him by poisoning with arsenic over a five years period. He died on the island of St. Helena in 1821; the official cause of death at the time was stomach cancer. In 2001, 180 years later, analyses of Napoleon's hair identified the true cause of death: arsenic poisoning.

Hair tells tales.

ADDITIONAL COMPONENTS OF THE HAIR FOLLICLE UNIT

So far, we have been discussing the hair follicle itself because it alone is responsible for the construction of the hair. There are two other components of what is called the pilosebaceous unit that are external to the hair follicle but directly connected to it. Associated with each hair follicle on the head is a sebaceous gland, which is important for maintenance of the hair shaft, and the erector pili muscle, which is, well, weird.

Sebaceous glands

The sebaceous glands are microscopic glands in the skin that are attached to the hair follicle. Sebaceous glands secrete sebum into the hair follicle to condition the hair on your head and the skin of your scalp. Sebum (Latin, meaning *fat* or *tallow*) is made of lipids, waxes and the debris of dead fat-producing cells. In the glands, sebum is produced within specialized cells and is released as these cells burst.

The function of sebum is to lubricate, to protect and to waterproof hair and scalp. Sebum keeps hairs from becoming dry, brittle and cracked by dehydration. The glands deposit sebum on the hairs, and the sebum rises to the skin surface along the emerging hair shaft. Sebum is the cause of some people experiencing oily hair in hot weather or if hair is not washed for several days.

The activity of the sebaceous glands increases during puberty because the glands enlarge in response to the elevated, circulating levels of androgens in the bloodstream. Teenage males can have very greasy hair. Women's hair tends not to be as greasy as men's hair because, in general, women produce less sebum than men do. As we age, we make less sebum and, as women age, their hair may become drier and duller because of the age-related decrease in activity of the sebaceous glands. Sebum itself is odorless, but when greasy hair is not washed often enough, the sebum is broken down by bacteria and can smell. Excessive production of sebum has been linked to eating red meats and fried or oily foods, but studies are not conclusive.

Erector pili muscles
One end of a tiny muscle, called the erector pili, attaches to the dermal layer of the skin, near where the hair shaft emerges, and the other end attaches deep down to a fibrous layer near the bottom of the hair follicle. Picture a tight sling from the skin to the base of the hair follicle.

The erector pili muscle is a type of muscle called smooth muscle or involuntary muscle. Smooth muscles are different than skeletal muscles as in your arms and legs. Typical smooth muscles work in your blood vessels and digestive tract. You cannot willfully control a smooth muscle as you can a skeletal muscle. You can think about moving your arms and legs, and they will move. You can think about not blushing or quieting a churning stomach, but it won't happen.

There are nerves that end in the erector pili muscle and when these nerves are stimulated, the erector pili muscle contracts and hairs stand up or, at least, try to stand up. The effect is small but, nevertheless, fright can make your hair stand on end. Some of the hairs on your body also have erector pili muscles and contraction of the erector pili in your skin gives you goose bumps. Goose bumps are even mentioned in the Bible. Job reports: "Fear came upon me and trembling...the hair of my flesh stood up." Why this happens when a person is scared, or whether there

HAIR

is functional significance to the erector pili muscles' ability to tug on hair follicles in the head and in the skin, is not obvious to anyone.

We have just visited the construction site for making a hair deep in the hair follicle. But different people have different types of hair. What is the basis for qualitative differences in hair and why do people have naturally curly, wavy or straight hair? Sooner or later it had to come up: genes.

4

TYPES OF HAIR: FROM TIGHTLY COILED TO STRAIGHT

"...eat the crust of the bread if you want curly hair."

- old wives' tale

Your haircut is important because it frames the features of your face. Your hair color is important because it complements your complexion. Your hairstyle is important because it makes a statement about your personality. But really, at the end of the day, it is all about the only things that you cannot change: the qualities of your hair. The qualities of your hair are related to the type of hair that you were born with.

In terms of natural appearance, unaltered by mixed genetics or by cosmetic hair treatments, there are three visually distinguishable types of hair that can be generally categorized by geographic origin: African, Asian and European. Categorizing natural types of hair based on the geographic origins of people allows us to discuss the inherent qualities and characteristics of hair: contour, shape, thickness, shininess, strength, as well as hair color.

Hair types are different because of small differences in genetic backgrounds. There are approximately 170 genes in the human genome that influence the shape, color and type of hair in all people. As humans evolved in specific geographic regions, slight changes in some of the genes related to hair have led to differences in hair types. There can be, to some extent, individual variations within each hair type, mostly in European type hair. And, there are exceptions. For example, some people from Europe have very straight hair, almost like Asian type hair. However, these variations are relatively small compared to the overall categorization of hair characteristics into types by geographical regions of origin. When people from different geographic regions mate, those hair genes get mixed in their offspring and the broad, regional classifications of hair types break down.

Within any of these geographic categories, there are no detectable differences between men and women in the qualities of hair. Comparing hairs from people from Africa, Asia and Europe, there are no differences in the building blocks for constructing hair. Furthermore, there are no differences in the keratins or the underlying biochemistry to make the building blocks. Nevertheless, there are obvious hair types in people of different geographic regions.

AFRICAN TYPE HAIR: THE FIRST HAIR

African type hair was the first kind of head hair to evolve, presumably from fur on the head, and is now representative of all native Africans who live south of the Saharan desert. Less than one million years ago in Africa, the sparse, short fur on the heads of hominids began to grow longer. Specific genes that are expressed in the hair follicles became modified and the changes in head hair brought by these variant genes were favored by the lifestyle and the environment of certain groups of our ancestral line. These slightly modified genes were naturally and/or sexually selected by evolution.

Perhaps the new hair on the head protected the scalp from UV exposure as early human-like ancestors walked upright in the intense African sunlight in regions around the Equator. But, this new type of hair did not grow very long. The relatively short lengths, to which African type head hair usually grows, combined with its springy coils, results in an airy, almost sponge-like effect. This may facilitate an increase in the circulation of cooler air onto the scalp. Think about having very short hair and wearing a loosely woven straw hat. Sweat would evaporate quickly and you would be cool in the sun.

African type hair has a relatively low water content, and does not respond like straight hair to moisture and sweat. African type hair has a slow rate of swelling or taking up of water, so it does not stick to the neck and scalp when wet. Unless totally immersed or drenched, African type hair tends to retain its basic springy texture. This characteristic, as well as its relatively short length, undoubtedly contributed to enhanced comfort levels in intense equatorial climates. In tropical climates, or after getting caught in the rain, European and Asian type hair tend to naturally fall over the ears and neck when wet. Very uncomfortable.

As our ancestors migrated out of Africa to populate the world, they had tightly coiled hair on their head that certainly looked different than fur. The coiled contour of the head hair on people from African origin is due to the shape of the individual hair shafts, which are different from hair shafts on the heads of people of European or Asian origin.

In cross section, African type hair has a ribbon-like shape; the hair shaft is flattened and asymmetric. When African type hair emerges from the narrow opening of the hair follicle, it grows almost parallel to the scalp and twists around itself as it lengthens to produce the characteristic tightly coiled hair. The ribbon-like shape of African type hair is produced by the lower end of the hair follicle. Deep in the hair follicle, in the bulb region, there is a bend in the hair shaft such that the hair shaft looks like a golf club. In the construction zone of the hair follicle, there is an asymmetry in the matrix of cells that are working to form the hair shaft. There are more keratinocytes and melanocytes on the convex side of the bulb (outer part of the head of the golf club) and fewer workers on the concave side of the bulb (inner part of the head of the golf club).

This asymmetrical distribution of cells in the hair follicle makes the ribbon-like hair shaft have thicker and thinner regions. It is this asymmetry of construction that gives these hair shafts their characteristic tight coils.

The shape of African type hair is genetically programmed into each and every hair follicle in the scalp. Remarkably, when hair follicles from people from African origin are excised and grown individually in a laboratory tissue culture dish, each hair follicle continues to produce a flat, asymmetrical, ribbon-like hair shaft. Similarly, when hair follicles from people from European origin are excised and grown individually in the laboratory, each hair follicle continues to produce an oval, symmetrical hair shaft. If African type hair follicles from the head of an individual were transplanted to the head of an Asian or European, what would happen? There are no reliable, accessible clinical records of transplanting African type hair to Asians or Asian type hair to Africans, but my guess would be that the hair shafts would continue to grow in their original shapes.

On a person of African origin, the appearance of the head of hair is less thick than the head of hair on Asians or Europeans. African type hair is the slowest growing of the three types of hair at, on average, less than half an inch per month or less than six inches per year. The hair follicles are significantly less dense on the heads of Africans than on the heads of Asians or Europeans. Therefore, the head of hair on a person of African origin generally appears less full than Asian or European type hair.

African type hair has a wiry texture and is often twisted. All along the hair shaft, there are natural constrictions, knots and fractures that cause the interlocking of several hair shafts that are bundled or bunched. African type hair produces plenty of protective sebum from the sebaceous glands, actually more sebum than European or Asian type hair. However, due to the tight curls, the sebum fails to spread evenly along the hair fiber and, without lubrication, the hair shafts become dry. These factors cause the brittle strands of hair to flake and roughen, resulting in hair that can become coarse and easily breakable.

TYPES OF HAIR: FROM TIGHTLY COILED TO STRAIGHT

Because of its inherent fragility, African type hair breaks off and does not easily grow as long as Asian or European type hair. The brittleness of African type hair underlies the belief that it cannot be grown long. Although African type hair does break more easily than Asian or European type hair, there are no underlying biochemical differences producing inherent weaknesses that account for the brittleness in African type hair compared to other types of hair. Because the tight curls create stresses at each turn in the hair fiber, the hair shafts become weak and fragile, making them prone to breakage. As a result, tightly coiled hair tends to naturally remain short. However, as seen at the local Reggae club, African type hair can indeed be grown long in the form of dreadlocks.

THE FIRST SELF-MADE WOMAN MILLIONAIRE
Sarah Breedlove was born in 1867, the daughter of former slaves. When she was 20 years old, she moved to St. Louis where three of her brothers were barbers. Like many women of that era who did not wash their hair often, she experienced hair loss. The young woman had an entrepreneurial spirit, experimented at home, and developed a shampoo and a sulfur containing scalp ointment. She traveled extensively throughout the south, door-to-door, promoting her products and supplying local African-American women with her hair lotions. Demand for the products took off and soon a national mail order business was established. The products were sold under the brand name Madame CJ Walker and the leading one was the ointment, marketed as Madam Walker's Wonderful Hair Grower. At the height of her company's success, Madame Walker had 3,000 highly trained door-to-door salespeople. She also established a college in Pittsburgh to train "hair culturists" in the use of her products.

Through the success of her business, Madame Walker became very wealthy. She was well recognized in the civil rights movement in the early 20th century as a philanthropist and as a speaker, giving empowerment lectures to African-American women on how to start their own businesses. At the time, Madame Walker was the richest African-American woman in the US and the first self-made, female millionaire. Living in her mansion, *Villa Lewaro*, on the Hudson River until her death in 1919, she was a neighbor of John D. Rockefeller. Madame Walker's home was often used for summits of leading figures in the civil rights movement.

TODAY'S CONSUMERS: AFRICAN-AMERICAN WOMEN (AND SOME MEN)

In 2009, Chris Rock, the TV comedian, produced a movie documentary on what African-American women do to their hair in pursuit of "*Good Hair*," which is the title of the movie. Through interviews and observations, Rock demonstrates how African-American women and men feel about their hair. In the film, many of those questioned expressed frustration and actual dislike for their hair in its natural state.

African type hair does not conform easily to specific hairstyles unless high heat and strong chemicals are used. Many African-American women want long, straight, shiny hair that swings to-and-fro when they move their heads. They want their hair to be more like European or Asian type hair. And, they will spend a lot of money and endure pain in the pursuit of good hair.

One chemical approach that some African-American women and men use is a relaxer. Relaxers can be applied in a salon by a hairdresser or at home. A solution of relaxer contains one of the harshest chemicals on earth, sodium hydroxide. The hair and scalp are soaked with a relaxer until the person's head feels like it is burning, and then the relaxer is washed out. Because the more the burn, the more relaxed or straight the hair will be, clients will go for the maximum burn that they can tolerate. Professional, medical articles in the dermatology field warn against the use of relaxers and highlight the damage, perhaps permanent, that they do. Nevertheless, the daughters of African-American

TYPES OF HAIR: FROM TIGHTLY COILED TO STRAIGHT

women are having their hair relaxed at very young ages. More on relaxers in Chapter 11.

A costly alternative for African-American women seeking good hair is weaving or extensions. In a salon, the natural hair is knotted close to the skin and a matrix is formed to which new hair can be attached or woven. The attached hair is an extension, and can be human or artificial hair in a variety of colors. Most of the human hair that is woven into African-American women's hair is from India. There are members of a Hindu sect in India that completely shave their heads, twice in a lifetime. This ritual is a self-sacrifice act of purification, called *tonsure*, and is carried out in a temple in Tirupati, India, where two million people per year get their hair completely shaved off. The shaved, long, Indian women's hair does not go for waste. It is gathered up by entrepreneurs, cleaned and made into long strands of hair or locks that are sold in the US hair products market. Many African-Americans are wearing Indian women's hair. In New York City's hair district and Los Angeles' fashion district, there are small, family-owned stores where a wide variety of imported hair for weaves can be purchased by wearers or hairdressers.

As talked about in Chris Rock's movie, an African-American woman who has a weave or an extension must be treated just so. Having the weave done can cost thousands of dollars, so women feel they have to protect their investment. No swimming, no showering without a shower cap, no sweating and, perhaps most important, no touching during sex. African-American men are sometimes wary of a woman with a weave. These women are considered high maintenance, but the women respond that they are just trying to have long, manageable hair, which is hard to grow for them. Of course, many in the African-American community choose a natural look: short, Afro-style or dreadlocks.

The high cost of these hair extensions has led to a new kind of burglary. Thieves are breaking into beauty salons, bypassing the cash register and safe, and stealing the high-end remy hair that is usually stored under lock-and-key. Remy hair is long, silky, human hair from India that has the outer cuticle intact making it is more easily shaped and styled, and less likely to tangle. Burglars are getting away with hundreds of thousands of dollars of human hair extensions that are then sold on the black market to hairstylists. During one recent break-in, the proprietor of the salon was shot dead. Salon owners are hiring guards, and installing surveillance systems and anti-theft devices.

Now, remember, African type head hair was the first head hair and undoubtedly evolved hundreds of thousands of years ago. Approximately 70,000 years ago, our ancestral lineage migrated out of Africa. About 40,000 years ago, hair began to change.

ASIAN TYPE HAIR: THE SECOND KIND OF HAIR

Asian type hair is the fastest growing type of hair at more than half an inch per month. In cross section, an Asian type hair shaft is thick and round, having the greatest diameter and area with circular geometry. Asian type hair is characteristically straight and has the most elasticity, strength and resistance to damage compared to African and European type hair. The mouths of Asian type hair follicles are round in cross section and the hair follicle is situated in such a way in the scalp as to cause the hair shaft to grow straight and almost perpendicular to the scalp. The density of hair follicles on the heads of Asians is, on average, greater than African but less than European densities. Therefore, Asian type hair does not appear to be as thick as European hair.

Asian type hair has a characteristic shine. The cuticle layers, the cross sectional shape of the hair shaft and the ability of sebum to coat the full length of the filament determine the amount of shine. Straight Asian type hair usually appears shinier than tightly coiled hair because of its smooth surface, allowing maximum light reflection and ease of sebum to move from the scalp down the hair shaft. Asians have such good hair because of variant genes.

Think of a gene as a long string of hundreds or thousands of beads, each bead having the letter A, G, C or T on it, in what seems like a random sequence. Actually, the sequence of letters is not random but is a specific code giving the cell instructions on what protein to synthesize. Now think about an almost identical string of beads lying next to the first one but with one of the beads having a changed letter, perhaps a G changed to an A. The second string of beads is like a *variant gene*. Often, the term *polymorphism* is used. The change in one letter in the code of the gene may cause the cell to make a slightly different, slightly abnormal protein. Or, the change in one letter of the gene may cause the cell to make more, or to make less, of the protein. We all have variant genes; they are what make each one of us different from each other. If there were no variant genes, we would all be clones and identical in every way.

Asian type hair evolved from African type hair about 40,000 years ago before people wandered into the eastern territories of Asia, India and eventually the South Pacific. At about that time, in a common group of migrants, there was a one letter change in at least two genes that are important for hair growth. One gene is the ectodysplasin A receptor, or EDAR, and the other gene is the fibroblast growth factor receptor, or FGFR. Human genetic studies have shown that these two variant genes occur only in Asians. The variant EDAR gene causes the specific protein that is made from that gene to have less activity than normal. The variant FGFR gene causes more of that specific protein to be made than normal. Somehow, and we do not know how, those changes result in hair with the round cross sectional shape and the straightness of Asian type hair.

The frequency and appearance of the variant EDAR and FGFR genes increased rapidly and widely in the Asian population those tens of thousands of years ago. Now, almost half of the world population has head hair with Asian type characteristics. The impact of such vast acceptance of these variant genes into the world population implies some huge evolutionary benefit to the group. But, what is this advantage? One could argue that long, straight hair protects the head against cold temperatures by preventing heat loss and that the thicker head of hair produced by these variant genes may have been advantageous as early people migrated into the cold climates of northern Asia. However, this explanation does not satisfy questions about why people of southern Asia, who live in warmer climates, have characteristic Asian hair. Admittedly, we do not know why slight variations in the EDAR and FGFR genes resulted in such a significant change in hair type.

Maybe these genetic variations have nothing to do with hair. An alternative explanation is that the changes in the EDAR and/or FGFR genes cause functional changes that affect another trait, not related to hair but, nonetheless, very important to Asian people. For example, the same variant EDAR gene also causes shovel-shaped teeth in Asian people. The shape of the teeth of Asians, particularly the front teeth, the incisors, is different than the teeth of people from Africa or Europe. In addition to the shovel-shape, there are prominent vertical ridges on the teeth. Thus, the variant EDAR gene causes at least two changes: round, straight hair and shovel-shaped teeth. Maybe it was the change to shovel-shaped teeth that was favored by evolution because of the new diets of these early migrating people and the straight hair came along with the change as a bystander effect. Once the new type of hair appeared as an evolutionary add-on, it was probably considered desirable and then met the forces of sexual selection. Specialized teeth and more attractive hair would make this genetic variant establish itself quickly and become widespread throughout the Asian populations. Whatever the explanation, a very large proportion of the people in the world have straight, black hair.

EUROPEAN TYPE HAIR: THE MOST DIVERSE HAIR

By far, European type hair has the most variability and diversity in what are considered the qualities of hair. European type hair grows, on average, slightly slower that Asian type hair at half an inch per month. Seen in cross section, European type hair is elliptical or oval in shape and has the lowest cross sectional area of the different hair types. The elliptical shape of European type hair contributes to its waves or curls. European type hair grows from hair follicles that have an ovoid shaped mouth at the surface of the skin and this type of hair grows at a slightly curved, oblique angle to the scalp, which also adds to the wavy quality. The density of hair follicles on the heads of Europeans is the highest of the three geographic groups.

Of people of European descent, approximately 45% have straight hair, 40% have wavy hair and 15% have naturally curly hair. When European type hair is naturally curly, the hair follicles growing these hair shafts are more like the golf club hair follicles of African type hair. The cross sectional shape of the hair shaft determines the ease of grooming. Friction caused by combing is lowest in straighter European hair, and straight or wavy European type hair allows easier and more styling options.

Long, straight, blonde hair, particularly from Eastern Europe, commands a price. As discussed above, there is a large market in the US for hair from India being used for weaves and hair extensions for African-American women. There is also a large market in the US and other affluent countries for blonde hair for weaves and hair extensions. The golden tresses come from rural Russians, Ukraines and women in other formerly Soviet bloc states. With the collapse of Communism, many of the already poor people in these regions are destitute. Growing crops is difficult, but growing long, blonde hair, harvesting it and selling it to intermediates representing international buyers brings some, temporary financial relief. Blonde hair is particularly preferred because it is easier to dye and match the color of the recipient's own head hair before it is woven into place.

The qualities of hair include volume, waviness, luster, strength and elasticity. A century ago, there were hundreds of "snake oil treatments" for improving the qualities of hair. Today, perhaps not much has changed. We are constantly exposed to advertisements from the hair care industry that offer innumerable products containing all kinds of substances that are claimed to improve the qualities of our hair.

HAIR THICKNESS AND VOLUME

In the mid 1800's, JC Ayer of Lowell, MA added Ayer's Hair Vigor to his marketing line of potions to treat throat and lung afflictions, scrofula, King's Evil, constipation and ague. Dr. Ayer owned an apothecary where he experimented and mixed up his patent medicines. Ayer's Hair Vigor was originally sold as a treatment to make hair thicker. The ingredients in Ayer's included lead acetate, cream of tartar, caustic soda and glycerin, and the product sold well into the 20th century. These were the good old days of quackery. Today, the beautiful peacock blue glass bottles and posters with women with voluminous hair are collector's items. New products have replaced the old potions.

People who have thick hair can have several things going for them. Hair thickness is dependent on the density of hair follicles, the size of the hair follicles and the thickness of the hair shafts. Of course, in a given individual any or all of these factors can contribute to thick hair.

You are likely to have thick hair if the density of hair follicles in your scalp is greater than 1600 hairs per square inch. With this high density the total number of hair follicles in your scalp is near the higher side of the range, 150,000. As indicated earlier, people of European origin tend to have some of the most voluminous head hair on the planet and that is because of the high density of hair follicles on their heads.

Hair thickness is also heavily influenced by whether the actual hair shaft is thin or thick. Thickness is determined inside the hair follicle and is a function of the size of the hair follicle. Smaller hair follicles produce thinner hair shafts; whereas, larger hair follicles produce thicker hair shafts. Hair follicles at a high density that produce thick hair shafts give the appearance of the fullest head of hair.

Contributing to hair thickness is the shape of the mouth of the hair follicle. The rounder the mouth of the hair follicle, the thicker is the hair shaft. The greater the square area of the mouth of the hair follicle, the thicker is the hair. Big hair follicles produce thicker hair shafts. Structurally, a thick hair shaft has more layers of cuticle, more keratin in the cortex and a larger medulla region. The architectural structure of all of these layers is controlled by genetics and no hair care product can change your genes.

HAIR WAVINESS AND CURLINESS

Most people who have curly hair wish that they had straight hair. Most people who have straight hair want wavy hair. In any event, a lot of resources, time, money and energy are put into making the transformation. You can be somewhat successful in reshaping the contours of your hair, at least temporarily, but the problem is, here too, you are going against your genes.

The TCHH gene, on chromosome one, is the major gene controlling the lengthwise contour of hair. The protein made from the instructions on the TCHH gene, trichohyalin, plays a role in the development of hair follicles in the embryo. In fully formed adult hair, the protein is a substantial constituent of the medulla, the center of the hair shaft. Normally, the TCHH gene instructs the synthesis of trichohyalin that makes the hair shaft relatively straight.

The natural waviness or curliness of hair is caused by slightly different forms of the trichohyalin protein that are inside the hair shaft structure. From human genetic studies, we know that slight variations of this protein, perhaps by as little as one amino acid change that is caused by a variant TCCH gene, can be responsible for the different, lengthwise physical contours of hair, making the hair wavy or curly in people of European origin. Nothing that can be bought at the drug store or the hair salon will change the TCCH gene or the protein, trichohyalin, made from the gene. But, there can be temporary fixes.

In addition to genes, the manner in which the 16 subtypes of keratin proteins are assembled into the protofibrils in order to build a hair shaft are likely to contribute to the waviness or curliness of hair. Comparing different people, even within a family, variations in contour and strength may be due to slightly different genetic programs using various combinations or proportions of hair keratins to form the cortex of the hair shaft. Depending on which keratins you are programmed to make and use, hair will take on different physical forms or properties. Also, your hair may feel different on different parts of your head. Within each person, slightly different proportions of certain keratins in hairs in the front of the head, compared to hairs in the back of the head, may be the reason that the quality of hair varies in different regions of your head.

HAIR LUSTER

Hair luster or shine is an optical phenomenon that is dependent on the biological properties of hair. The luster of hair is a component of attractiveness, thought to imply healthy hair and, therefore, a healthy body. Think of the head of a person with long, dark straight hair that catches the light in such a way that a shining band appears across the hair. That is luster.

Coating the external surface of the hair shaft, the cuticle contains millions of small, flat scales that reflect light. The scales are in layers. The higher the number of layers, the higher the amount of reflected light and, therefore, the higher the luster.

Luster of the hair is most pronounced in black Asian type hair. Compared to European type hair, Asian type hair has wider cuticle scales and more cuticle layers. Also, the cuticles in Asian type hair are heartier and stand up to stress better than European type hair. More layers of the shingle-like cells, which are packed more closely together, give Asian type hair the smooth surface and better sebum coating needed for high shine. The shine coming off long, black Asian type hair is hard to beat. Generally, curly hair does not shine.

Luster of the hair can be negatively affected by damage to the cuticle of the hair shaft as occurs with artificial hair coloring and styling of the hair. Also, frequent shampooing can remove too much sebum and affect shine. On the other hand, dirty hair due to excessive sebum accumulation lacks luster.

HAIR STRENGTH AND ELASTICITY

It may not be obvious when handling a single hair, but you only need to try to break a small lock of cut hair by pulling on the ends to be convinced that hair is extremely strong. The organization of keratin within the cortex of the hair shaft allows a hair to resist a strain of up to 100 grams (one quarter of a pound). If you glued one end of 100 long, cut hairs to a 20 pound weight, you could bundle up the other ends of the cut hairs and lift the weight. Perhaps there is some truth to the cartoon depicting a caveman pulling a woman by her hair back to his cave. Theoretically, the average head of hair could hold 12 tons of weight, if the person was hanging upside down and the scalp were strong enough.

Before breaking, a hair undergoes changes. If you can pluck a 12 inches long hair from your head or from a friend's head, delicately and slowly pull from both ends of the hair. The hair shaft behaves like an elastic thread; after extending slightly, when released it returns to its original length. Hair has elastic properties and hair scientists have instruments to measure such properties of hair. Used in the cosmetic industry to test the effectiveness of hair care products, an extensiometer slowly and progressively stretches a hair, allowing a precise study of the modifications that hair undergoes before it breaks. The elastic properties of hair, due to the structure of the keratin molecules, permit stretching during which hair can be pulled 30-50% longer. Keratin in the natural, coiled state progressively rearranges itself into keratin in a non-coiled state and begins to resist. When the stretching stops, the hair shaft returns to its initial form like a spring. That is elasticity. If the stretching continues to 50-60% longer than the original length, a condition known as flowing occurs in which the hair shaft elongates but will not return to its original length. Elasticity is lost. At 70-80% stretch beyond its original length, the hair shaft will break.

When stretched beyond the elastic phase, hair has another property that lasts for a while, plasticity. The hair shaft keeps the new length and contour that it has been given. Thus, if a hair is stretched and wound around a curler and later the curler is removed, the hair retains the new curliness or waviness. In combination with water and heat, this plastic property allows temporary modifications of the contours of hair.

The properties of strength and elasticity vary greatly depending on the type of hair. African type hair breaks under a strain of 50 grams after an elongation of 40% and is the most fragile. At the other end of the scale, Asian type hair is the strongest and is elastic up to an elongation of 55% under a strain of 100 grams. For both of these qualities, European type hair has an intermediate position. However, by extrapolating these various measurements from different types of hair to theoretically equal shapes and surface areas, scientists have shown that hair from these three regional groups behave in an intrinsically comparable fashion. Such findings further demonstrate a common, basic, underlying structure for all the hair in the world.

WET HAIR

After your shower, your wet hair is obviously heavier than when your hair was dry. This simple observation illustrates an important characteristic of hair: hair is permeable. Despite the close fitting scales of its cuticle and the waxy sebum that naturally coats it, a hair in good condition can absorb more than 30% of its own weight of water. When you wet your hair, your hair takes up a lot of water. When you use shampoo, your hair takes up even more water. Furthermore, if the hair is already damaged by other factors, the percentage of its own weight of water absorbed can reach 45% and the diameter of the hair can increase by 15-20%. Hair coloring products cause swelling of the hair to enhance penetration of large dye molecules into the body of the strand of hair.

In general, water is harmful to hair and considerably amplifies many of the environmental damaging factors. Water particularly accentuates the negative effects of sunlight. When you return from a vacation of basking in the sun and swimming in the water, your hair has been traumatized. While you were enjoying the pleasures of letting your hair dry in the sun after bathing in the sea or pool, you were in fact producing significant structural damage to the hair shafts. Excessive water and sun altered the keratins, making the hair fragile and more easily damaged. In addition, with intense and prolonged exposure of wet hair to sun, hair becomes lighter because of de-coloration due to degradation of melanin. Fortunately, the construction zone of the hair follicle is not damaged and will eventually grow out your original hair structure and color. But that will take time. At a hair growth rate of six inches per year, by the time your hair grows back, you will certainly be ready for your next vacation.

Hair takes up water even without bathing in the sea, taking a shower or shampooing. Hair is permeable to water vapor, which is always present in the surrounding air to a greater or lesser extent. The presence of this water vapor is known as relative humidity. Ah, you are already frowning. Water taken up by your hair from the atmosphere when there is relatively high humidity in the air around you is why you have a bad hair day during humid weather.

ELECTRIC HAIR

The contact of hair with the material of certain natural or synthetic fabrics can produce sparks. When you take off a pullover sweater, your hairs stand up on your head. Similarly, a plastic ruler rubbed on a piece of material will attract hair. These nuisances are associated with the ability of hair to become charged with static electricity.

Friction gives hair an electrical charge that causes individual hair shafts to repel each other. Wet hair has higher combing friction than dry hair, but wet hair conducts electrical charges well so the charges are dissipated. Dry hair conducts electrical charges poorly so when combed or rubbed by a wooly sweater, the charges build up, produce static electricity and each hair shaft does not want to be near any other hair shaft. Damaged hair makes things even worse. The ability to accumulate static electricity increases when the cuticle is stressed, encouraging even more electrical charges to associate with the hair shaft. In dry weather, a long time spent in the shower shampooing cuticle cells away can also cause a bad hair day.

FETAL HAIR

During human embryological development, the fetus goes through a furry stage, lanugo. By 22 weeks in the uterus, a developing human fetus has all of its five million hair follicles formed for life and is covered with fine downy hair that resembles sparse, thin fur. The lanugo hair is the first hair to be produced by the hair follicles and is usually lost to the amniotic fluid by 33-36 weeks of gestation. However, babies born prematurely may emerge covered with lanugo hair all over their body or on portions of their body. When a recent preemie was born with a beard, his empathetic dad showed solidarity and family resemblance by not shaving until his son came home from the hospital, which was three months later and without a beard. Interestingly, lanugo hairs can develop on the arms and chest of women patients with anorexia nervosa.

VELLUS HAIR

Vellus hair replaces lanugo hair and grows from the same hair follicles in the developing fetus in the uterus. Why there are two cycles of hair growth in the human fetus is unknown. In people, vellus hair is the vestigial hair that is similar to fur on an animal. The fine, short hair shafts that cover over most of the human body are colloquially referred to as peach fuzz. Most vellus hairs are barely noticeable. They are non-pigmented or lightly colored, and grow at a low density. Vellus hairs grow to much less than an inch long, drop out, and are replaced by another vellus hair. Thus, vellus hairs do not continue to grow long, as does head hair. The hair follicles that produce vellus hairs on the body are not associated with sebaceous glands, as are the hair follicles of the head.

Both males and females have vellus hairs. On children and adult women, vellus hair is most easily observed on the arms and legs. In males at the time of puberty, hair follicles producing vellus hairs are turned into hair follicles producing terminal hairs on the body by the actions of androgens, released from the gonads and circulating in the bloodstream.

When hormones start flowing in men, vellus hair is replaced by thicker, longer, pigmented body hair on the chest, arms, underarms, legs and beard. Women undergo these changes on their legs and underarms during puberty. Adult men generally have more and darker, terminal hair on their body than women. Pubic hair also arises during puberty from vellus hairs in both men and women. Transiently, vellus hairs appear in the balding areas of men and women who have androgenetic alopecia (Chapter 7).

BABY HAIR
Babies may be born with a full head of hair, or not. In newborns, all of the hair follicles on the head are there, but they are producing either vellus hairs or fine, thin terminal hairs. Each baby is different and, when born, the hair follicles on the head can be at different stages of development. As the baby grows, the vellus producing hair follicles on the scalp convert to producing terminal hairs and the hair follicles grow bigger and, therefore, make thicker hairs. The activities of melanocytes, which make the color pigments for the hair, are also developing and changing in the growing baby. Therefore, the thickness and the color of the hair of the newborn will often change as the baby grows, and the hair follicles develop into their mature role of providing the child and adult head with hair. As you get old, as we all know, that head of hair will change again.

AGING HAIR
With age, hair usually becomes less thick and the overall capacity to produce long hairs on the top of the head is reduced. As there are age-related changes in all organs of the body, there are age-related changes in the hair follicles. On an old head, there is a decrease in the density of the hair follicles. Furthermore, of the remaining hair follicles, there is a decrease in the number of active hair follicles in the anagen growth phase and an increase in the number of hair follicles in the telogen resting

phase of the hair growth cycle. In addition, many of the remaining hair follicles miniaturize and produce thinner hair strands of poorer quality. Also, there is probably a lifetime of cumulative damage to the hair follicles from personal and environmental insults. Undoubtedly, some of the sebaceous glands gave out a long time ago. With age, the thickness and fullness that you were so proud of when you were young is lost. Well, haircuts are less frequent with age.

Although all hair is made the same, biologically, the three different types of human head hair, African, Asian and European, have different qualities. The greatest diversity of hair qualities is in people from European origin. In terms of color, most of the head hair of the people of the world is naturally the same color, black or a very dark brown. However, there are different natural colors of head hair in people of European origin. Over tens of thousands of years, there were several changes in the hair of the heads of European people. In addition to waviness and curliness, changes in natural hair color were also occurring in this relatively small population on the earth. Let's examine what gives hair its natural color.

5

NATURAL HAIR COLOR -OR- WHY ARE THERE REDHEADS?

"Redheads are said to be children of the moon, thwarted by the sun and addicted to sex and sugar."

-Tom Robbins, *Still Life with Woodpecker*

With the advent of modern, relatively easy hair treatments and dyes, the color of hair can be any color of the rainbow: red, orange, yellow, green, blue, indigo or violet. But, you were not born with these colors in your hair, so let's get down to the natural roots.

The natural color of hair can have a varying palette, but not a rainbow: black, brown, auburn, red, strawberry-blonde, blonde, gray, silver and white. All of these colors are due to the relative amounts and types of a special group of chemicals, pigments, which are collectively known as melanins. As described in Chapter 3, melanins are put into the base of the growing hair shaft deep inside the construction zone. Because hair keeps growing, a month after hair is artificially colored, there is about one half inch of natural color emerging from the hair follicle. The roots are showing.

One of the unappreciated aspects about hair color, when you stop to think about it, is that almost everyone in the world has hair that is naturally a very dark color. Greater than 90% of the world population has black or dark brown hair. There is in fact not great diversity in natural hair colors around the globe. For example, you will never see a person whose ancestry is from New Delhi with naturally red hair, or a person from Nairobi with naturally blonde hair. Essentially, with very few exceptions, only people of European origin, a relatively small population, have a large variation in natural hair colors, from black to blonde. It is a matter of chemistry.

PIGMENTS ARE COLORFUL CHEMICALS
Pigments are chemicals, natural or man-made, that have a color. They are unusual, because not all chemicals have a color that we can see. There are also minerals that have color, for example gold, silver, copper and carbon. These and other natural chemicals were used as the pigments in paints by artists throughout the ages. Today, complex chemical structures that are made in laboratories and factories form the bases for the colors in paint, plastic, ink and dyes.

There are natural pigments that are made by living things all around us. The green color of leaves is due to the pigment chlorophyll and the red of our blood is due to the pigment hemoglobin. The brownish-green shell of a live lobster is due to dark pigments that are destroyed by cooking, leaving heat-resistant carotenoid pigments which miraculously transform the color of the shell to red. Our world is full of colors from natural pigments and our eyes have evolved to see an infinite variety of these colors.

Melanins, which are the pigments that are produced by our bodies, give color to hair, eyes and skin. Although scientists have tried to find it, there is no strong relationship between hair color, eye color and skin color. Many people have dark brown hair, brown eyes and brown skin. However, many people have brown hair, blue eyes and light skin. And, of course you have seen brunettes, blondes and redheads with a variety of skin colors and brown, blue or green eyes. The synthesis of melanins in the hair, eyes and skin uses the same genes, but these three organs are genetically programmed differently and are not coordinated in any way that produces a color link between them.

MELANINS GIVE HAIR ITS NATURAL COLOR

Melanins are a class of chemical polymers made up of many repeated and linked molecular units. Polymers are made from small chemicals that are joined together in a repeating manner to make bigger, longer, larger and more complex chemicals. For example, processed rubber and plastic are complex chemical polymers made in factories. Starch in plants and glycogen in your body are complex carbohydrate polymers. Melanin polymers are like long chains that might be used to hold down a load on the back of a truck.

Synthesis of melanin polymers, melanogenesis, goes through a series of steps and needs the instructions from several genes. The enzymes and structural proteins that are made according to those instructions participate in the making and the packaging of the melanins. There are two basic types of melanin polymers in human hair: eumelanin and pheomelanin.

Eumelanin actually has two subtypes: black eumelanin and brown eumelanin. These are black or dark brown insoluble nitrogenous pigments derived from oxidative polymerization of repeating subunits of 5, 6-dihydoxyindoles, which are made from the amino acid tyrosine. Chemically, pheomelanin differs significantly from eumelanin. Pheomelanin is a yellow/red alkali soluble, sulfurous pigment derived from oxidative polymerization of repeating subunits of cysteinyldopas. The eumelanin and pheomelanin polymers may be further chemically modified by melanocytes to give subtle differences in color. Not that you need to know all of this.

The natural color of your hair is due to eumelanin and pheomelanin inside the cortex and the medulla of the hair shaft. The amounts of eumelanin and pheomelanin put into a hair shaft is controlled by genes, and the variations in hair color among individuals are due to having more or less eumelanin, and more or less pheomelanin in each person's hair shafts. Generally, if more eumelanin is present and less pheomelanin, the color of the hair is darker; if less eumelanin is present and more pheomelanin, the hair is a lighter color. The relative amounts of the types of melanins can vary over time causing a person's hair color to change. Also, it is possible to have hair shafts of more than one color and hair shafts in different parts of the head can be a range of related colors.

When chemicals are used to lighten hair or hair is bleached by the sun, the amounts of eumelanin and pheomelanin in the hair shafts are being altered. Eumelanin is less chemically stable than pheomelanin. Therefore, the breakdown of eumelanin is faster during hair coloring or under intense sunlight, resulting in relatively more pheomelanin remaining in the hair and making the hair appear lighter in color. Eventually, the pheomelanin will also breakdown and the hair will gradually become orange, then yellow, and finally white. That is why hair coloring treatments are timed; don't let that hair lightening stuff stay on too long. More on coloring hair in Chapter 11 when we go to the hair salon.

THE WORK OF MELANOCYTES

Melanocytes are the highly specialized cells that make melanins. The melanocytes in the skin produce melanins that absorb UV radiation and protect the surface of the body from the damaging effects of sunlight. Without melanin in the skin (albinism), a person would fry to a crisp during a day at the beach.

The melanins made by the melanocytes are placed in little bags called melanosomes. People with dark skin have more melanocytes, melanosomes and melanin than people with light skin. Anthropologists tell us that when our species migrated out of Africa into northern latitudes where there was decreased intensity of sunlight, lesser amounts of melanin in the skin was needed to maintain the natural benefit of sunlight, activating vitamin D. So, our skin gradually evolved to be lighter. Interestingly, as skin color became lighter, hair color remained dark in our distant ancestors.

As discussed in Chapter 3, the hair shaft is growing during the anagen phase of the hair growth cycle because keratinocytes are making keratin and lengthening the basic scaffold. Keeping up with the nonstop activities of the keratinocytes are about 100 melanocytes per hair follicle. Every day they must produce more melanins, package the melanin polymers in melanosomes and transfer the melanosomes to the keratinocytes at the base of the hair follicle. Melanocytes in the construction zone of the hair follicles work hand-to-hand with keratinocytes, performing dangerous and demanding work that eventually leads to their depletion, as we will see in Chapter 6 on graying.

THE DIVERSITY OF HAIR COLORS

Black is the natural hair color of most people around the world. This dominant genetic trait can be found in people of many backgrounds, ethnicities and geographical regions of the earth. Black hair is the darkest and, by far, the most common of all human hair colors. The black color is due to the presence of large quantities of black eumelanin in the hair shaft, and can be perceived as soft black or blue black. In English literature, black hair is sometimes described as jet black or raven black.

Black hair has certain qualities, for example, it is the shiniest of all hair colors. Some scientists argue that it is not possible to have black hair. Instead, they describe people as having very deep, dark brown hair, which is seen as black because the large amount of eumelanin in the hair gives it lustrous properties that make it appear black. If you are in Asia, look carefully at the hair of the people you meet. Black or very dark brown?

Brown is one of the natural colors of hair in people of European origin. Brown hair is due to the presence of large quantities of brown eumelanin in the hair shaft and is the next most common hair color, varying from light brown to almost black. Brown hair is common in the United States, and central and southern Europe. Brown-haired people can also be found in western Asia and northern Africa. The strands of brown hair are usually thicker than those of light colored hair but not as thick as those of red hair. Brown-haired people are known as brunettes, who are often thought of as being smart and serious. Some brunettes do not like the color of their hair, referring to it as "mousey brown." George Bernard Shaw may have been first to describe lackluster brown hair when Professor Higgins refers to Eliza Doolittle's hair in *Pygmallion*: "Her hair needs washing rather badly: its mousey color can hardly be natural."

Auburn hair contains a varying mixture of high levels of brown eumelanin and low amounts of pheomelanin, and can be described as a color that is somewhere between brown and red. Always the social critic, Mark Twain wrote in *A Connecticut Yankee in King Arthur's Court:* "When red-headed people are above a certain social grade, their hair is auburn." Auburn hair ranges from light to reddish brown and is most common among people of northern and western European descent.

Red hair ranges from vivid strawberry shades to deep auburn and burgundy. Red hair has the highest amounts of pheomelanin and usually very low levels of brown eumelanin. The range of the variation of shades of red hair is caused by slightly more or slightly less brown eumelanin. Between 1 and 4% of the world population has red hair, depending how one defines red. Red hair occurs more frequently (up to 6-10% of the population) in England, Ireland and Australia. In the US, about 2% of the population has red hair. Obviously, all of these people were from European descent. Interestingly, red hair has been found on ancient mummies discovered in the Tarim basin of northwestern China. The people who lived there had the features of Caucasians, not Asians, and probably migrated from the lands of Europe some 4,000 years ago.

NATURAL HAIR COLOR -OR- WHY ARE THERE REDHEADS?

There is more on the subject of red hair than the relative proportions of pheomelanin and eumelanin. People with red hair are special, so special that they are given their own section below.

Strawberry-blonde hair contains a mixture of pheomelanin and brown eumelanin. Strawberry blond is thought to have originated in Celtic and Scandinavian countries, and is an even rarer type of hair color than red hair.

Blonde hair contains small amounts of brown eumelanin and no, or very little, pheomelanin. Blonde hair ranges from nearly white (platinum blonde) to a dark golden blonde. A little more pheomelanin creates a golden blonde color, and a little more brown eumelanin creates an ash blonde.

Less than 2% of the world population has naturally blonde hair which occurs more frequently in women than in men. Blonde hair is seen in people of European origin, but is rare among people of non-European origin. Based on genetic studies, people with blonde hair became numerous in northern Europe several thousand years ago, thus blonde is a relatively new hair color to the human species. The wide ranging entry of blondes into the human population and history was undoubtedly due to sexual selection. After all, gentlemen prefer blondes.

There are a few isolated pockets of people with lighter color hair amongst general populations of people with black hair. Light brown hair is found among the Yukhagir of eastern Siberia and some Inuit bands of the western Canadian Arctic have fair hair. Auburn colored hair is sometimes seen among the indigenous people of Taiwan. Blonde hair has arisen independently among some Aborigines of central Australia and the Solomon Islands. These are obviously genetic anomalies and are kept within the indigenous populations because of their isolation.

Changes in hair color occur in young children as they grow. Parents of children born with lighter hair colors may find that the hair of their kids gradually darkens as their children get older. Many blonde, strawberry-blonde, light brown, or red haired infants experience a gradual darkening of their hair into adulthood because a higher percentage of brown eumelanin is added to the hair shafts as the child matures. Why this happens is not known and there is nothing that can be done, currently, to stabilize the child's light color hair.

REDHEADS ARE SPECIAL

Throughout history, people with red hair have been noticed and cultural reactions to red hair have varied from ridicule to admiration. Redheads have been considered to be hot-tempered, seductive, sexually great lovers (the women, not the men), evil, not trustworthy, fearsome, liars and witches. Lilith, the original, mythical woman on whom Eve and the story of creation was based, is often described as a redhead. It is believed that Judas was a redhead. There are great female redheaded warriors in Irish myths and Vikings that stand out, for example Erik the Red. In medieval times, Jews were often portrayed as redheads, like Shylock in Shakespeare's *The Merchant of Venice*. In Marion Roach's book, *The Roots of Desire: the myth, meaning and sexual power of red hair*, the admittedly redheaded author traces back in history a poisonous potion that required the "fat of a red-haired man." The book goes on from there with all sorts of crimson musings and is well worth reading if you are interested in red hair.

Doctors know that redheads can be different. At Christmas time 2010, the *British Medical Journal* ran a feature about redheaded patients. Many surgeons and anesthesiologists believe that certain precautions must be taken when a redhead is undergoing surgery. In fact, there is medical evidence that extra general anesthesia may be needed when operating on a redhead because of their decreased pain threshold. Also, redheads seem to need more local anesthesia. Picture this; don't do this. If you had a redhead and a brunette strapped to armchairs and gradually started pinching their forearms with increasing strength, the redhead would say "ouch" before the brunette. In a large population based study in the US, natural redheads were found to have twice the risk of developing Parkinson's Disease, compared to dark haired individuals. Some

NATURAL HAIR COLOR -OR- WHY ARE THERE REDHEADS?

doctors believe that redheads have a tendency to bleed more. This may or may not be true but redheads do bruise more easily. Furthermore, there are some medical findings that redheads form hernias more often than dark haired people. Interestingly, there is an association between red hair and a rare eye disease, brittle cornea syndrome. Medically, redheads are special people.

Red hair is caused by certain variants of the melanocortin receptor (MC1R) gene on chromosome 16. Normally, stimulation of the melanocortin receptors cause the melanocytes to make certain amounts of brown eumelanin. When one or two variant MC1R genes are present in a person's genome, the genetic instructions produce a slight change in the melanocortin receptor proteins. These variant melanocortin receptors in people with red hair do not work as well as the normal receptors and in some cases do not work at all. Therefore, the melanocytes in the hair follicles of people with variant MC1R genes make much less eumelanin and in fiery red redheads, make no eumelanin. The less eumelanin made and, therefore, the more pheomelanin present in the hair shaft, the redder is the hair. Interestingly, variants of the MC1R gene only occur in people of European descent. Variants of this gene are never found in people of African or Asian origin.

The variant melanocortin receptor gene entered the gene pool of a group of humans that was living in areas of the northern British Isles about 40,000 years ago. In this neighborhood, way back then, our *Homo sapiens* ancestors were killing off that other, earlier species of hominids, the Neanderthals. The Neanderthals lived on the land that our ancestors wanted and hunted the animals that they liked to eat. Fortunately, Neanderthals were not as smart or as fast as our ancestors were. Before the total demise of the Neanderthal line, there was probably a lot of inter-subspecies sex. The Neanderthal genome shows that these early relatives had the variant gene for red hair. Did the variant gene for red hair enter our genetic history through the progeny of *Homo sapiens*-Neanderthal mating? Perhaps redheads have a little Neanderthal in them.

The persistence of the variant gene for red hair in the genetic pool suggests that the variant offers some advantage to the individual and the population. The evolutionary advantage may not be red hair per se. Certainly, the melanocortin receptors perform several functions. The MC1R protein is present in other cells in the body, not only the melanocytes in the hair follicle. Red hair may be a side effect of some advantageous genetic change that is significant, not to the melanocytes of the hair follicle, but to other cells and tissues. For example, the melanocortin receptor is an important gateway for immunological responses. But if the variant MC1R gene leads to a less functional melanocortin receptor, would that benefit immunological responses? Clinically, redheads have been reported to have different immunological responses and different sensitivities to pharmacological treatments with drugs. In addition, red hair is often associated with certain skin conditions: fair color, freckles, and high sensitivity to the UV rays in sunlight. Nevertheless, the possible advantages to the functions of other cells in the body of having the variant MC1R gene are unknown. Until we know what evolutionary advantage goes along with having red hair, the reason why redheads appeared in the European population will not be clearly understood.

Redheads are on the increase. Women who are not redheads are using artificial means to make their hair red. In today's world, many actresses have dyed their hair red, many elderly women favor dying their hair red and there are women of Asian and African descent dying their hair red. Obviously, these women think that red hair makes them more attractive. There are not many men, famous, elderly or otherwise, dying their hair red. What is it about redheads?

GENETICS OF HAIR COLOR
This subject is extremely complex because at least 14 genes are associated with the spectrum of natural hair colors from blonde to black. It is unknown how these genes are regulated to produce more, or less, eumelanin and pheomelanin in melanocytes in the hair follicle. We do

know that fiery redheads have the variant MC1R and that this gene dominates over other genes that program hair color. The real mystery of hair color genetics is why these variant genes that produce different hair colors are present in people of European background but essentially never show up in African or Asian people.

EVOLUTIONARY ADVANTAGE OF HAIR COLOR

According to the Hollywood stereotype, blonde women are beautiful and youthful, and redheaded women are hot and seductive. Certainly, we can at least agree that they are different in appearance than the 90% of the women in the world who have dark hair. When the traits of red hair and then blonde hair first started appearing during the prehistory of the northern lands of what we now call Europe, some primitive, dark haired guys and gals must have gone wild. There can be no doubt that the exotic looking redheaded and blonde women and men were in demand for sex. Sexual activity eventually led to a lot more redheads and blondes amongst the population. This is sexual selection and it is a strong force in Darwin's theory of evolution. OK, but why does hair have any color in the first place?

Some scientists have suggested that because melanins absorb UV radiation, long strands of hair containing pigment on top of the skull may protect the skin from the scorching effects of the sun. Sunburns only occur in regions where there is no hair; ask any bald man who forgot to rub on SPF 45. But wait a minute. Don't buy into this misconception that hair has melanin to protect your head from the sun. You can question that idea with something that you may already be aware of. People with a full head of gray or white hair (much less or no melanin, as will be explained in the next chapter) do not get sunburned scalps. Therefore, the presence of melanins in the hair as a protection against sun damage does not really fit with what we see on people. Protection from the sun is not why there are pigments in hair.

Other scientists have other thoughts. A possible reason for having melanins in hair is that it acts as a waste depository. One of the properties of melanin is that it binds and adsorbs all kinds of foreign chemicals, which is why we have telltale hair. For example, melanin can bind many man-made chemicals and heavy metals that are toxic. Binding of toxic substances to melanin and then transferring the melanin to the growing hair shaft effectively removes the toxic materials from the bloodstream where they could affect important organs. The toxic materials cannot be reabsorbed from the melanins in the hair shaft because the hair shaft is growing away from the blood vessels in the scalp. Then, a haircut will forever rid the body of these toxins. In this hypothesis, melanin is key in the hazardous substance handling described in Chapter 3 as a possible reason for having long, continuously growing hair. Are melanins in hair a part of toxic waste management for the body and that is why hair has color? Seems like a rather elaborate explanation.

Another suggested reason for having hair color is the process of melanogenesis itself. If we are to have hair, then the rich densities of hair follicles on our heads are potential entry points for microorganisms to gain access into our bodies. The biochemical processes of melanogenesis generate all kinds of free radicals, like hydrogen peroxide, that can kill these microorganisms. Do we have hair with natural colors so that the processes of producing melanins kill germs that may enter the hair follicles? Seems like a rather convoluted explanation.

We are born with hair with color but end up with hair that does not have color. So, why does hair have color in the first place? When an aging adult forms gray hairs, the pigments in hair are no longer being made (as you will learn in the next chapter). There are age-related diseases but they are not due to the absence of pigments in gray hair. Without more scientific input, this section ends with a big: Who knows why people have black, brown, red or blonde hair?

NATURAL HAIR COLOR -or- WHY ARE THERE REDHEADS?

CHANGES IN HAIR COLOR AFTER DEATH

Being dead may make some people go red. Hair lasts longer than dead flesh because hair was already dead when the person died. Remember that eumelanin is less chemically stable than pheomelanin and breaks down faster when oxidized. And remember, the less eumelanin, the redder the hair. Because of the oxidation of eumelanin over millennia, well-preserved Egyptian mummies appear to have reddish hair. They were dead long enough under the right conditions. The color of hair changes slowly under dry oxidizing conditions, such as burials in sand. In the damp conditions inside a wood, metal or plaster coffin, hair does not last long enough to change color.

In this chapter, we have noted that almost everyone in the world has dark hair. For some reason, a small segment of the world population, the early Europeans, started sprouting hair that was not black but ranged in color from brown to red to blonde. Did this give Europeans some advantage in making wine and cheese, or was colorful hair just considered exotic and sexy? If the latter, why didn't variations in hair color pop up in other regions of the world? In any event, there is one natural hair color that we all end up with: gray.

6

THE INEVITABLE GRAYING OF THE HEADS OF THE WORLD

"Aged? But he does not appear aged, just look, his hair has remained young!"

-Marcel Proust, Remembrance of Things Past

The medical term for hair getting old, *canities*, is the age-dependent change in natural hair color to gray and eventually white. The change occurs gradually over years or decades. Graying is slow because each hair follicle is doing its own thing either producing a hair with natural color or a gray hair. A hair follicle that is turning out a gray hair may be right next to a hair follicle turning out a brunette hair. However, once graying begins, each year more and more hair follicles join in to produce gray hairs and white hairs. When a hair follicle starts making a gray hair, it will never go back to making a hair with the original, natural color ever again.

You can be lucky and have CEO hair, that smooth, shining, combed mane of hair that is always perfectly swept back in place and seems to gleam along with capped teeth. But most people do not have the CEO look. Gray hair is coarser, wirier and less manageable than naturally colored hair. An extreme example is the Albert Einstein look. In some people, gray hair is unable to hold a permanent or even a temporary set, and it is harder to artificially color gray hair. Gray hair shafts tend to be thicker and grow faster than normally pigmented hairs. They defiantly point out in different directions and can form odd curls, spirals and stick-like random shapes.

Graying is another of nature's enigmas. It is part of our being human and it affects an obviously visible part of our body. There are no other mammals that gray to the extent that we do. The causes of graying are known and, in the grand design of things, could have easily been avoided. However, genetic information that causes graying was not selected out as we evolved. Canities was retained in the human genome as a physiological change in the human body and we all gray as we age. One can only conclude that graying has a purpose. But, what is that purpose?

Gray hair is an early, easily visible sign of aging. The onset of graying occurs after the peak of mating activities when sexual abilities start to decline. Graying may be a signal alerting a young, prospective mate that the potential object of their affection has peaked physically. Gray hair appearing on your temples may interfere with your ability to attract a younger mate in their sexual prime. On the other hand, white, silver, gray, and certainly salt-and-pepper hair can be attractive the second time around. Fortunately, graying does not usually produce strong feelings of loss of self-image, as balding does. Perhaps gray hair is a marker of the time for building more mature relationships. Most people with gray hair are seen as more settled and down to earth. The partying is over.

LIKE AN OVERWORKED PRINTER
Let's use as an analogy, a printer, perhaps a home printer, that contains a black ink cartridge. By this analogy, the printer is the hair follicle,

the cartridge is a melanocyte and the blank page is a strand of hair. You learned in Chapters 3 and 5 that melanocytes are the cells that produce the melanin pigments that give hair its natural color. Like a melanocyte, the printer cartridge contains pigments, in this case black ink. A message sent from your computer programs the printer to make letters and the cartridge is activated to release pigment onto the plain white paper moving through the printer. Throughout most of the life of the cartridge, pigment is released and the page is printed with crisply defined letters. Towards the end of the life of the cartridge, when there is not much pigment left in the cartridge, the letters on the page become lighter and lighter, actually grayer. Further printing causes fainter print on the page. When there is no ink left in the cartridge, the page running through the printer will come out completely white.

An analogous process to the printer cartridge running out of pigment is happening in the melanocytes of the hair follicles as one ages. Gray hair is like a printed page with very light lettering and white hair is like a white paper that goes through the printer but emerges without any printed letters on it. The melanocytes, like the cartridge, are spent and can no longer perform their function of supplying melanins to color hair. For the printer, you can go out and buy a new cartridge. However, you cannot shop for replacement melanocytes on the Internet. But, the hair follicle has an intrinsic replacement strategy. Spent melanocytes are replaced from a pool of melanocyte stem cells within each hair follicle. This strategy works…for a while.

If you run the printer constantly, day and night, to print out pages of text, the cartridge will soon run out of pigment. Perhaps you can get a few days of constant printing out of a cartridge. Remarkably, the melanocytes can go on producing melanins and adding it to the hair shaft every day for years. However, the ability to maintain natural hair color throughout life faces two impeding designs in the system: melanocytes work with dangerous materials and the number of replacement stem cells is limited.

A CLOSE LOOK AT GRAY HAIR

As you did in Chapter 2, put on your explorer's hat and let's do some searching through a jungle of hair. Don't shrink down just yet. You will need a colleague in this exploration, a friendly consenting adult, about 40 years old, with hair that is at least 4-6 inches long, preferably even longer. Now, run your fingers through your friend's hair and look up close until you find a hair with color that fades to gray at the root. With his or her permission, pull it out. By examining the hair shaft, you can clearly see that the graying of the hair begins at its root, regardless of whether the hair has its original natural color or has been dyed.

If your partner in exploration has natural hair color, and has not dyed their hair, go back to his/her head and find a hair that is mostly gray. Promising this will be the last time, pull it out. You should be able to see a gradation of gray visible to the bare eye, or you may need the magnifying glass on the Explorer version of your Swiss Army knife. Starting at the tip of the hair, there may be almost natural color but, as you scan down to the root of the hair, the hair becomes gray, then lighter gray and may be even white at the root. Hair does not become gray by all of the pigment falling out all along the hair shaft. Hair becomes gray as it grows.

Now, let's shrink ourselves down again and examine what is going on at the root of a hair, in the construction zone where a hair shaft is made and colored.

LIFELONG OCCUPATIONAL HAZARDS

The graying of hair is all about the melanocytes or, more specifically, the untimely loss of function and the death of the melanocytes. Melanocytes work in a dangerous microenvironment. During melanogenesis, lots of nasty chemicals known as free radicals and reactive oxygen species are continuously formed while the synthesis of melanin is ongoing. These chemical byproducts are hazardous to health. Free radicals and reactive oxygen species are often blamed for the aging process, and are well known to damage cells and tissues in which they are formed. Many of the dietary supplements that we take, those jars just below where the pharmacist stands, are supposed to prevent free radicals and reactive oxygen species from forming or neutralize them if they do form. None of them prevent gray hair.

The formation of these highly reactive, dangerous substances is a necessary and inherent part of the chemical processes of melanin synthesis. Now that we are down here in the construction zone, you can ask the melanocyte: "If you work in a dangerous environment, don't you have any protection?" The melanocytes do have the cellular machinery, anti-oxidant chemicals and enzymes to protect themselves from being damaged by the free radicals and reactive oxygen species in the immediate environment. However, the ability of the melanocytes to protect themselves does not last the lifetime of the hair follicle in the head.

For example, as one of the byproducts of synthesizing melanin, melanocytes make hydrogen peroxide, which is a free radical and a very hazardous chemical. Hydrogen peroxide is that liquid in a brown plastic bottle that you buy in the drug store and fizzes when you pour it on an open cut. It kills germs and almost everything else. Melanocytes in hair follicles that are constructing gray hairs have very high levels of this reactive substance. The melanocytes normally make catalase, an enzyme that renders hydrogen peroxide harmless. However, the aging hair follicle does not produce sufficient catalase to handle the abundance of hydrogen peroxide. The high levels of hydrogen peroxide in the aged hair follicle damage the enzymes that form melanin and may, in addition, directly bleach some of the melanin in the hair shaft in the construction zone.

For the first 30 years or so, the melanocytes are doing well. But, after 5-10 hair growth cycles, the melanocytes are becoming tired, damaged and depleted, and they are no longer able to synthesize melanin as they did in their youth. The environment of free radicals and reactive oxygen species in which they work damages the melanocytes and they slow down and die off. In the hair follicles producing gray hairs, the melanocytes appear vacuolated and clearly contain less pigment than that of a young person.

With poorly functioning, and fewer and fewer melanocytes, a smaller number of melanosomes containing less melanin are transferred into the cortical keratinocytes of the growing hair shaft. When the melanin content becomes low enough, the growing hair shaft takes on the appearance of a gray color. Eventually, with the complete absence of melanin, melanosomes and melanocytes in the hair follicles, the growing hair shaft becomes white. It is intriguing that death of the melanocytes occurs at an early age, even though the hair follicle continues making the actual hair shaft into old age. If a hair follicle loses the ability to make a hair with natural color at age 30, that hair follicle will still go on making hair for another 40-50 years. Whether you have a full head of hair or you are bald, gray or white, hair is made until you die.

Graying occurs because of the occupational hazards in the life of the melanocytes working in the highly reactive construction zone of the hair follicle. There is a backup plan to replace the melanocytes, but the backup plan lacks longevity.

THE SHORT LIFE OF STEM CELLS

At the end of anagen, the hair growing part of the hair growth cycle, the hair follicle takes a rest. While it rests, the construction zone is dismantled. Many of the melanocytes that were so active in the coloring of hair during anagen die off. They are expendable and most will not be used in the next hair growth cycle. Lacking a construction zone, the hair follicle shrinks to one third of its original size. But, the rest period is relatively short, about three months. Going back to work means rebuilding the construction zone and, when you are young, the stem cells in the hair follicle are there to help.

Dormant stem cells reside in a small bulge or niche of the hair follicle below the skin but some distance from the construction zone at the base of the hair follicle. In the niche, stem cells wait for the call to action. After telogen, the resting phase of the hair growth cycle, the stem cells are activated to divide and to differentiate into melanocytes that will replace their spent brethren at the base of the hair shaft. As the stem cells migrate downwards into the newly rebuilt construction zone, they become melanocytes and happily contribute pigment to the hair shaft in the next active phase of hair growth.

However, there is not an infinite number of stem cells sitting around the niche in the hair follicle. At about age 30, when melanocytes are having trouble adequately protecting themselves from their hazardous work conditions, the stem cell pool is becoming depleted. The ability of the stem cells to maintain the population of melanocytes deep in the hair follicles decreases with the years. In the construction zone, hair is still being made, but the graying process is underway.

The limited number of stem cells that can potentially become melanocytes and the age-related, decreased functioning of melanocytes are responsible for the onset and progression of graying. Why is this happening at such a relatively early age? One reason is that we live longer than we were designed or evolved to live. It is important to understand that many of our genes are still operating on the assumption that we are cave-people who will only live to be about 30 years old. Age-related changes in certain body functions happen to everyone well before the person would be considered old. Another example of a significant functional change occurring early in adult life is presbyopia, the loss of near vision and the need for reading glasses. Apparently, these systems have not been reprogrammed to match our new lifespan.

Perhaps there aren't enough stem cells to last our whole life because melanocytes can be bad actors. A highly active and replicating cell like the melanocyte stem cell has the potential to become a threat to its neighbors or other parts of the body. Dedifferentiated melanocytes can turn cancerous and form melanomas, a dangerous and aggressive form of skin cancer. With age, we become more susceptible to this type of cancer. The early loss of stem cells of the hair follicles may be a protective mechanism against cell transformation into uncontrollable, wildly growing cancerous cells that become destructive.

THE FIRST GRAY HAIRS AND THEIR FOLLOWERS

The age of onset of graying is somewhat earlier in people of European origin than in people of Asian origin. The onset of graying in people of African origin is a bit later. The first gray hairs in Europeans appear, on average, by 30 years of age and the age of onset is similar in males and females. Hair is considered prematurely gray if onset is before the age of 20 in Europeans, before 25 in Asians and before 30 in Africans. Nevertheless, white hairs can appear as early as childhood. Sometimes, even babies are born with gray hair, but that is due to inheriting an unusual trait.

Neither the age of onset nor the progression of graying is dependent on the initial natural color of the hair. Brunettes, redheads and blondes all gray at about the same rate. However, graying appears to occur earlier in dark haired people than in redheads or blondes, because a few gray hairs will stand out more distinctly in people with dark hair compared to people with light hair. On the other hand, fair haired people appear to be completely gray at an earlier age than brunettes. The remaining natural colored hairs in a fair haired person, who has turned mostly gray, will blend in more easily with the graying hairs and the head of hair will appear all gray.

How old will you be when those failing melanocytes begin to let your natural hair color start slipping towards gray? How quickly will graying progress? As with many aspects of hair, it is a matter of genetics. You can take a hint from your parents: premature gray, early gray, late gray. You will probably follow the age of onset and the progression of graying of your mother or father. But it may be hard to get specifics from your mom.

The progression of graying is compounded because, with age, more hair follicles remain for longer periods of time in the telogen phase and, therefore are not actively growing hair. As the head of hair thins with age, the greater percentage of gray hairs is more apparent. In general for the US population, the graying of hair follows the 50/50/50 rule. Roughly, by age 50, 50% of people have 50% of their hairs that are gray.

There usually is a pattern to graying. In men, gray hair first appears at the temples and then spreads to the top of the head and the remainder of the scalp, with the side regions usually graying last. In women, gray hair usually starts around the perimeter of the hairline. Why graying shows these patterns has no apparent explanation. But, one notes, that the pattern of graying is similar to the pattern of balding. However, there is no causal or any other association between graying and balding.

THE FALL OF THE PIGMENT

When the leaves of trees lose their primary pigment, chlorophyll, in the fall, they lose their green color. What we see in the annual fall display is the underlying color of the leaf material that can be yellow, orange or red and which was always been there but hidden by the presence of the green pigment.

When hair starts to lose its melanins, and then progressively loses more and more of its pigments, we see various shades of gray in an admixture of hair shafts with different amounts of pigmentation. But, gray hair is not actually gray; it is yellow. This underlying color is due to the presence of keratin, the structural protein that remains in the hair shaft as less and less melanin pigments are put into the growing end. The intrinsic color of hair keratin is light yellow. The appearance of hair as gray is an optical effect; the reflection and refraction of incident light by aged hair masks the yellow color and hair appears to our eyes as gray when it still contains low amounts of pigment. Sometimes, gray or white hair can have a yellowish tinge due to keratin. Occasionally, you will see a senior citizen with light gray or white hair that looks a bit yellow.

The manifestation of a single gray hair can occur in several ways. A single hair follicle can produce a single hair shaft that is grayer and grayer as it grows over time. In this hair shaft, there will appear to be a gradual loss of pigment starting at the distal end with the most natural color, and having less and less natural color as the hair shaft gets closer to the scalp. Such a hair shaft is being formed by a hair follicle that is losing its functional melanocytes during a single anagen growth phase over several years. Or, a single hair follicle can grow a series of hair shafts over many years and each subsequent hair shaft appears grayer and grayer. In this case, the hair follicle is losing more and more melanocyte function during repeated catagen phases and has fewer functional melanocytes as it is resurrected into the next anagen phase.

HIDING THE GRAY

The reader is warned that no product available anywhere can stop, slow or reverse the inherent biological processes causing gray hair. Remember Ayer's Hair Tonic, which was mentioned in Chapter 4 as a snake oil potion, advertised in the late 1800's, which could make hair fuller and thicker? Ayer's Hair Tonic was also advertised as a potion to stop the graying process. Fast forward to now and we have the latest products that claim to restore hair to its youthful look. By skipping over the last 150 years, I have not reviewed a large number of hair products, mostly containing Chinese herbs, which were often advertised at the back of magazines. One of the currently available products is Melancor, an over-the-counter pill that claims to prevent graying and restore gray hair to its natural color, as well as prevent baldness. From the advertising, Melancor can "visibly revive dead hair cells." Quite a claim. The product needs no prescription and has not been approved by the FDA. Results, meaning no more gray hair, are expected after swallowing these pills for 6-12 months, "usually." Two tablets taken twice a day for 12 months will cost about $400. Worth a try?

Gray hair can disappear in a variety of ways. You can pull out each gray hair as they appear. No, it is not true that if you pull out one gray hair, two gray hairs will grow in its place. Or, you can cover them up chemically. Temporary hair colorants are large organic molecules that do not penetrate the cuticle, the surface of the hair shaft. The colors are not intense and wash out with subsequent shampooing. Henna is a natural hair colorant that can add red highlights but on gray hair can look quite orange.

Most effective at covering gray are permanent hair dyes. These dyes penetrate into the hair shafts and are activated by a chemical treatment. The color from permanent hair dyes withstands normal hair washing; however, as the dyed gray hair grows, the roots need to be touched up. More on coloring hair in Chapter 11.

There are gradual or progressive dyes like a certain formula of doubtful Grecian origin. These are usually marketed to men and contain lead acetate, a toxic chemical that is added to gasoline. As the coloring solution is rubbed on the hair, it penetrates through the cuticle into the cortex. You will remember that the cortex of the hair shaft contains sulfur containing keratin proteins that are all bound together by disulfide bridges. The lead ions in the gray hair dye react with the sulfur atoms in the proteins to form lead sulfide, which happens to be a dark color. After lead acetate treatment, the once gray hair looks darker and a man formerly with gray hair may appear more youthful. The appearance is adjustable. The more frequently the solution is applied, the darker the "lead head" becomes.

A LITTLE PATCH OF GRAY

Piebaldism is a rare autosomal genetic disorder of melanocyte development. One of the characteristics of this disease is a congenital forelock of white hair on the forehead. Piebaldism is inheritable.

Poliosis is a patch of gray hair or a white streak that can appear anywhere on the head of healthy children and adults. This patch of gray can be inherited and is sometimes associated with skin disorders or other pathologies. Occasionally, babies are born without melanocytes in the hair follicles in a small region of the scalp. In some cases, the immune system may inadvertently destroy melanocytes in this region.

MARIE ANTOINETTE SYNDROME

This colorfully named medical entity is the condition in which hair suddenly turns white after extreme stress. There are numerous historical anecdotes of hair turning suddenly white; the time course has been reported to be overnight or over several days.

As the story goes, Marie Antoinette's hair, which was a great source of pride to her, suddenly turned white the night before her execution in front of the Paris mobs in 1783 by that monument to French efficiency, the guillotine. However, there were several earlier reports of Marie Antoinette's hair turning white during her attempted escape to flee France and subsequent capture, and upon hearing of the King's untimely demise. Maybe the hair of Marie Antoinette turned white not overnight but over a short time period while the French were revolting. Or, maybe she took her wig off.

Earlier claims of hair suddenly turning white exist. There is the story of a scholar being appointed head of the Israeli Talmudic Society at a very young age. According to his wife, he developed 18 rows of white hair. The Jewish philosopher Maimonides suggested that this was the result of laborious studying.

The hair of other leaders as well as civilians has reportedly whitened quickly. Sir Thomas More suddenly turned white in the Tower of London the night before his execution (1535). In his epic poem about the battle, Marmion (1808), Sir Walter Scott writes: "For deadly fear can time outgo, and blanch at once the hair." There were reports of people's hair turning white during the incessant bombing attacks in England during World War II. Even in the medical literature, there are occasional reports of Marie Antoinette Syndrome.

However, there is no medical or biological basis for Marie Antoinette Syndrome. Most docs just don't believe it. As recently as 2009, there was an editorial in *Archives of Dermatology*, published by the American Medical Association, stating that Marie Antoinette Syndrome does not exist. The best explanation that the experts have suggests the sudden falling out of normally colored hair, due to some autoimmune process, which leaves in place the white hairs that were already there. In other words, if on the morning of their executions they arose with white hair, there should have been a lot of pigmented hairs on the pillows of Sir Thomas More and Marie Antoinette. If they slept that night.

It is hard to imagine the entire length of the melanin containing hair shaft suddenly becoming white. That would mean that all of the tiny melanin pigments would dislodge and fall out all along the hair shaft. Melanin polymers are embedded in the hair shaft and it is not possible for melanin to fall out through the cuticle. Melanin cannot just disappear overnight from already formed hair shafts.

As mentioned above, there have been suggestions, but not much known, about a possible autoimmune process that selectively attacks hair follicles that are producing hair shafts containing melanins. These autoimmune activities would have to cause the rapid loss of only hairs with melanins and not affect the white hairs already present. If the autoimmune process attacked the melanocytes present in pigmented hair follicles, perhaps inflammation would make the hair with color fall out. Don't know.

Marie Antoinette Syndrome remains an unexplained, folkloric myth and an unsolved, medical mystery.

HAIR

As if graying isn't enough of an insult to our desire to remain and to look youthful, there is another impact of our genetic heritage that for many is extremely hurtful to their self-image. Going bald. In the next chapter, the loss of hair as a person goes bald will be described. As with graying, we wonder why the genetic anomalies that produce baldness have persisted in human evolution.

7

BALDING (HIS AND HERS)

"God made a few good heads…and the rest he gave hair."

-unknown, clever bald man

Baldness is the state of lacking hair where it should be growing on the head. The most common form of baldness in men is the progressive, top of the head, hair-thinning condition called male pattern baldness or, medically, androgenetic alopecia. There is also a female counterpart, referred to in the past as alopecia. Although the causes of baldness in men and women were thought to be different, current concepts have blurred the lines of distinction and suggests that men and women are suffering from similar causes. We now refer to both men and women as having androgenetic alopecia. The American Academy of Dermatology estimates 80 million men and women in the United States have this type of baldness. In addition, both men and women have gradual thinning of their hair as they get older, but this is a different process than that which leads to shiny, bowling ball heads rimmed by a horseshoe of hair.

Baldness occurs in all kinds of people with all kinds of hair types. Going bald is not due to the type of hair that you have, the density of hair on your head, the rate of growth of your hair, the color of your hair, your style of life or your gender. Baldness is undoubtedly due to variant genes that interfere with the work of the hair follicles, but the story is actually quite complex and incomplete. Functionally, why people go bald is anybody's guess. What we know about baldness just doesn't make sense.

SOCIAL PERCEPTIONS

Nobody likes going bald. Young men losing their hair often feel mildly to moderately anxious, preoccupied, depressed, distressed, self-conscious and dissatisfied about their appearance and their body image. Most men worry about baldness making them look older. In the past, some desperate men let the hair on one side of their head grow longer and did a comb-over. Picture Rudi Guiliani, Mayor of NYC in the 1980's. With age and further hair loss, the feelings can become more intense and lead to a loss of self-confidence. If balding is happening to you and you want to commiserate, there is live streaming of balding distress stories and even an internationally syndicated radio/talk show at the website < thebaldtruth.com>. For women losing their hair, all of these feelings are substantially more intense and there can be significant negative psychological effects related to self-esteem and sexuality. Women are less able to cope with losing their hair than most men. Men and women going bald believe that everyone notices – and everyone does.

The reaction of the noticing public does not help the stressed out baldee. Bald men may be viewed by women as less attractive and older than men with a full head of hair. Women who lose their hair may be thought by the public to be unhealthy (for example, undergoing cancer chemotherapy), unattractive and old. Throughout time, men and women have shaved their heads as a fashion, religious, submissive, philosophical or political statement. Nevertheless, there is no getting around that the ideal is a full head of hair.

There have been numerous social studies on male pattern baldness. Interestingly, although bald men are thought of as older, they are perceived as more mature, more socially competent and more dominant, in a non-threatening way. However, young women tend to think that bald men are less sexually attractive and less sexually active. If you are an old, bald codger who still wants to do it, you may have to work harder at being charming in order to secure a lover or a mate.

DEFINITIONS AND PREVALENCE

As mentioned above, clinicians and researchers who specialize in the field of hair loss refer to the condition as androgenetic alopecia and apply this term to both men and women. The general belief is that this type of baldness is due to over stimulation of the hair follicles by androgens. Environmental or external factors do not influence androgenetic alopecia. "Androgenetic" refers to the condition being caused by both androgens and genes; "alopecia" is the medical term for loss of hair. In men, this condition is called: male pattern baldness; in women this condition is called: female pattern hair loss. In the scientific and medical literature, initials are often used, AGA: androgenetic alopecia; MPB: male pattern baldness; and FPHL: female pattern hair loss. However, as we will see, referring to androgens and genes as a straightforward explanation of hair loss has significant, questionable and unexplained issues. Also, most of the medical research on androgenetic alopecia has been done on men. Female pattern hair loss still needs lots more research. Accordingly, most of what follows in this chapter relates to male pattern baldness.

Dermatologists in the clinic classify male pattern baldness in stages on the Hamilton-Norwood scale: I-VII. The doctor compares what he/she sees on a male patient's head to a series of typical, graphic images illustrating the stepwise temporal progression of the balding head, from receding hairline to the end stage horseshoe. Because the sequence and the areas affected are different in balding women, the Ludwig scale is used and the graphic images of female heads progress to show the pattern of hair loss in women with female pattern hair loss.

Although baldness occurs in people throughout the world, the prevalence of male pattern baldness varies in geographic regions. Men from European descent have the highest prevalence of baldness and the 50/50/50 rule seems to apply. These men have a roughly 50% chance of losing 50% of their hair by age 50. Female pattern hair loss in the US usually begins at an older age than male pattern baldness. With age, the prevalence of mid-frontal hair loss increases, affecting 60% of women aged 80 and over.

Not every population has as much androgenetic alopecia as men from European origin. The prevalence of baldness in men of Asian origin is 20-30%, which is significantly less than in Europeans. In men of African origin, androgenetic alopecia occurs at an even lower frequency, perhaps 10%, compared to European men and Asian men at similar ages. Of course, as these populations mix and have children, these differences in prevalence will be blurred.

MINIATURIZATION OF THE HAIR FOLLICLE

All of the changes that lead to baldness occur in the hair follicles. As male pattern baldness progresses, there is miniaturization of the hair follicles, meaning that they become shorter, narrower and smaller. The working parts of each hair follicle are present for a while, but on a greatly reduced scale. Smaller hair follicles make smaller hairs that grow more slowly.

Through the process of miniaturization, the width of the hair shaft progressively decreases until scalp hair resembles vellus hair. Finally, the hair follicles atrophy in the balding region and are non-existent. Those hair follicles are gone, baby, gone. No treatment yet has made the hair follicles come back once they are gone. The best you can do is slow down the miniaturization process with drugs, discussed below, or transplant new hair follicles into the barren areas.

ANDROGENS

According to the androgenetic hypothesis, androgens are over stimulating the hair follicles in the scalp in genetically susceptible individuals. This overstimulation is what leads to miniaturization of the hair follicles on the head and, eventually, irreversible loss of hair and baldness.

Androgens are a category of steroid-like hormones, namely testosterone and dihydrotestosterone, which are made by the body, circulate in the bloodstream and are generally associated with maleness. But, androgens are not gender exclusive. Both men and women make androgens, which have important functions in both sexes.

Androgens are good. For men, they are primarily made in the gonads and released into the blood to act on target tissues in the body. The time of androgens' greatest accomplishments is puberty and indeed androgen levels in the bloodstream are then very high. When the male of our species reaches puberty, circulating androgens are raging and make him think about nothing but sex. Androgens in the bloodstream drive the development of maleness into the 20's and fuel the search for the perfect mate.

During puberty in males, androgens stimulate increased muscle mass as men become hunks, development of the testes into sexual machines, thickening of the vocal chords that lowers the voice, and hair follicles in the chin, cheeks, neck, pubic region, chest, limbs and underarms to produce dark, long, terminal hairs instead of peach fuzz. In women going through puberty, androgens are also driving the development of pubic and underarm hair, although at a younger age and at a lower levels than in men and with less rage.

Please note that the hair follicles on a young pubescent man or woman's head are *not* targets for androgens. During male puberty when that annoying, wispy, teenage moustache starts to grow, the hair on the head is not stimulated to grow more or faster or thicker. In males and females coming of age, androgens convert hair follicles making vellus hair into hair follicles making body hair, but have *no effect* on the hair follicles that are present in the scalp of the young man or woman. Another observation about puberty is that when androgens stimulate hair follicles, hair grows; hair follicles do not miniaturize and die. Tuck this pubescent information away for now because there is a riddle coming up.

An association of male sex hormones with hair growth has been observed for ages. In the early days of the Italian opera, when women were not allowed to appear onstage for a performance, castration (an awful way to decrease circulating male sex hormones) was promulgated in young boys to assure their contralto voices and their feminine appearance with no beard growth. From contemporary medicine, we know that women who have extensive hair growth on their face and body (hirsutism) are often effectively treated with drugs that block the actions of male sex hormones. Furthermore, androgen-like drugs (for example, clinically used testosterone) produce facial hair when given to women and stimulate body hair growth in men. For all of these reasons and many more, it was long believed that androgens, circulating in the bloodstream, were responsible for driving male pattern baldness. But, if you read the last two paragraphs again, you will realize that these observations relate to androgens causing beard or body hair to grow – *not hair on the head*. Stay with me now.

Androgens may be associated with androgenetic alopecia, but not androgens being carried to the hair follicles through the bloodstream. Male pattern baldness occurs at an age when circulating testosterone is declining or already low. Also, women with female pattern hair loss do not have high levels of androgens in their bloodstream. For many years it has been known that the circulating blood levels of androgens are not what matters to balding.

The androgens that cause male pattern baldness are putatively made in the hair follicle and have their effects in the same hair follicle. Recent research points to locally made testosterone, synthesized in the hair follicle and converted into the more active form, dihydrotestosterone, which is associated with androgenetic alopecia. In other words, the affected hair follicles that will miniaturize make their own testosterone and dihydroxytestosterone. The small quantities made locally by the hair follicles do not enter the bloodstream in appreciable amounts to raise the blood levels, but are sufficient to presumably stimulate androgen driven activities in what must be androgen sensitive scalp hair follicles. Also, there are reportedly high levels of the enzyme converting testosterone to dihydrotestosterone in the miniaturized hair follicles of the balding region of people with androgenetic alopecia. Getting technical about male pattern baldness, it is postulated that the local synthesis of excessive dihydroxytestosterone, not blood levels, is driving the processes unfolding in androgenetic alopecia.

But nothing in human biology is simple and androgens made in the hair follicles are not the whole story. There is a lot more complexity to male pattern baldness because, as you may know, there are genes involved.

THE GENETICS OF ANDROGENETIC ALOPECIA

There are several variant genes that must be present in an individual in order for baldness to occur. Let's examine a variation in the gene for the androgen receptor on the X-chromosome that appears very important. Discovery of this variant gene made headlines in the New York Times in 2006, heralding that the cure for baldness was not far away. The variant gene does not change the protein made from the instructions in the DNA but may increase the number of androgen receptors in each cell. More receptors usually cause the cell to work harder. Perhaps male pattern baldness is due to overstimulation of the hair follicles of the head by the androgens made within the hair follicle as well as the increased number of androgen receptors caused by the variant gene.

But still, the connection between androgens and balding is tenuous. Here is the riddle I promised earlier. I will call it: "The Androgen Paradox."

*How can overstimulation of the hair follicles by local androgens cause balding, when we know from puberty (1) that the hair follicles in the scalp are **not normally sensitive** to androgens and (2) that in other parts of the body androgens **stimulate** hair growth, not hair loss?*

This riddle undermines the concept that androgens and genetics cause androgenetic alopecia. To answer this riddle we have to hypothesize that the hair follicles in the scalp ***become*** sensitive, indeed hypersensitive, to androgens in men who will go bald, even though they were not sensitive during puberty. How can that happen? Good question.

Maybe there is a hint, more variant genes. The gene variant of the androgen receptor does not act alone to produce androgenetic alopecia. There must be more going on, on a bald man's head. Variants of other genes are apparently also necessary, perhaps for conferring sensitivity of a head hair follicle to androgens. Any one gene variant cannot produce baldness itself, but combinations of slight variations of other genes (two but probably three) may sensitize the hair follicles to respond to androgens. Several gene variants have been associated with baldness to date and are listed below. However, we do not actually know how these variants could make the head hair follicles sensitive to androgens, and cause miniaturization, atrophy and loss of the functioning hair follicles.

Here are the variant genes that may contribute to androgenetic alopecia:
- variants of the androgen receptor gene (X-chromosome) associated with male sex hormone stimulation
- variants of the EDAR gene (X-chromosome) associated with hair thickness
- variants of the P2RY5 gene (chromosome 13) associated with hair structure and texture
- variants of the SOX21 gene (chromosome 13) associated with anchoring the hair shaft in the hair follicle, the synthesis of keratins and the formation of the cuticle
- variants of the APCDD1 gene on chromosome 18
- unidentified variant gene on chromosome 3
- unidentified variant gene on chromosome 20

We first met the EDAR gene in Chapter 4. This gene can have several variations. One variation causes curly hair, one variation causes wavy hair, one variation causes straight hair and one variation may be involved in causing no hair, baldness. It would be great if we knew what function the EDAR gene has that affects hair in so many ways. Maybe someday science will explain all of this.

Baldness is often associated, in family conversations, with the maternal grandfather and, in fact, there is a relationship between the bald maternal grandfather and the bald grandson. Note from the above list that the male child who will go on to male pattern baldness may receive two variant genes on the X-chromosome from his mother. His mother received that X-chromosome from her father. In an evolutionary context this makes sense. If you are a mother, wouldn't you want your son to have the virility of your father?

The genetic basis for inheritance of baldness from the maternal grandfather is not an absolute. In many families with a bald maternal grandfather and two grandsons, one grandson is bald and the other grandson has a full head of hair. The mother of those two boys has two X-chromosomes, one from her father and one from her mother. Only one is passed on to a son. Thus, there is about a 50-50 chance that the X-chromosome from the mother containing one or more variant genes will be passed on to her son. If the son inherits the X-chromosome from his mother that does not contain the variant genes, baldness will not likely be in his future.

As listed above, there are several variant genes on other chromosomes that have been associated with baldness and could be inherited from the mother or the father. Undoubtedly, more variant genes will be found. Therefore, it is not just the gene variant passed on from the maternal grandfather through the mother that causes baldness. Inheritance of several of these variant genes from both the mother and the father contribute to their offspring's likelihood of androgenetic alopecia.

But even if hair follicles in the scalp have become sensitive to androgens, death of hair follicles due to overstimulation by androgens does not occur anywhere else in the body. If you have some thoughts about The Androgen Paradox relating to baldness, please contact me.

WHY IS THERE A PATTERN?

There is a well-known pattern to hair loss in androgenetic alopecia that is due to the loss of hair follicles only in specific regions of the scalp. Men always lose their hair starting with a receding hairline at the temples. Receding hairlines are usually noticed in males in their early 20's but can be seen as early as late teens. As the hairline continues to recede, an additional bald patch develops on top of the head. Over time, the receding hairline and the bald spot on top of the head merge and, with further hair loss, result in the familiar U-shaped or horseshoe pattern of remaining hair on the sides and in the back of the head. In androgenetic alopecia, the central regions of the scalp go bald leaving the peripheral regions with continuously growing hair.

The patterns of hair loss are different in men and women. Women lose some hair first in the frontal hairline but there is much greater loss all over their scalp. Typically the frontal hairline is not completely lost but the density of hair on the top and sides of the head is decreased, sometimes drastically, in women.

The pattern in men is so familiar but why is there a specific pattern? Why doesn't all of the hair fall out? Why does the progression of hair loss stop at well-demarcated places that make this pattern ubiquitous? The pattern is likely due to regional differences in the scalp but this hardly explains what is going on. Does androgen sensitivity of the hair follicles only develop in the balding areas on the top of the head? Do the hair follicles on the side and back of the head not become sensitive to the effects of androgens? In truth, we cannot explain how or why hair follicles in only certain regions of the scalp die, while other hair follicles in the same scalp keep right on growing hairs. More fuel for The Androgen Paradox.

AN ALL-ENCOMPASSING HYPOTHESIS?

A hypothesis regarding androgenetic alopecia appeared in *Medical Hypotheses*, a recognized journal with a long history of publishing hypotheses with little or no supporting data. The author, E. Tuncay

Ustuner, took on the issue of the shape of the remaining hair in male pattern baldness and female pattern hair loss, and also the question regarding the involvement of androgens. The hypothesis centers on the weight of the scalp and gravity.

The hypothesis goes something like this. The weight of the scalp is felt more on the top of the head, as compared to the sides of the head, and, with age, there is compression of the soft, top tissue between the scalp and the skull. Compression of this tissue hinders the hair follicles to enlarge downward into the skin during anagen and, therefore, the hair follicles on the top of the head cannot produce hair. The hair follicles on the side of the head bare relatively less weight and the supple thickness of the scalp remains. The male-female balding difference in pattern of the remaining hair lies in the difference in the shape of the skulls; presumably the force of gravity to compress the soft tissue is manifested differently in men and women. The hypothesis goes further to suggest that androgens are not the causative agents but rather a response that the body makes to try to stimulate the weakened hair follicles to grow hairs.

This hypothesis does make sense, to some extent. Compression of the soft tissue between the scalp and skull could be important to the loss of hair follicles with age. The compression may be due to gravity and/or may be a function of aging in certain, genetically predisposed people. Not everyone will suffer from androgenetic alopecia because the predisposition is due to the structure of the tissue, which, of course, would be controlled by the genes of the individual. Furthermore, one can imagine a regional effect on hair follicles such that there is less compression of the soft tissue on the side of the head, so that hair is not lost in this region. The shape of the skull is known to be different in men and women; perhaps the soft tissue has a different geometry. Finally, postulating that the increase in androgens and their receptors appears as an attempt to ***stimulate*** the weakened hair follicles is perfectly consistent with androgens stimulating hair growth in other areas on the body.

HAIR

The key points made by this all-encompassing hypothesis are very interesting, but data is needed. Anyone doing research on this hypothesis out there?

EVOLUTIONARY REASONS FOR MALE PATTERN BALDNESS

Do bald men have an evolutionary advantage? The prevalence of male pattern baldness is so high in men of European origin, that it must have some evolutionary implication. Male pattern baldness developed in our prehistoric ancestors, probably first appearing in men migrating with their tribes and families westward into the lands of what are now Europe, approximately 30,000-40,000 years ago. Baldness must have been favored by evolution because there had to be several genetic changes in these men that led to hair loss. Getting two, three or more genes to change in order to develop a new characteristic or trait is very unusual and very difficult. There has to be a reason.

There are examples of a single variant gene being favored by evolutionary pressure because of a specific advantage. The classic example is sickle cell anemia. In people with sickle cell anemia, the gene that gives the instructions for making hemoglobin is slightly different than the normal gene. This variant gene instructs the hemoglobin producing cells to make an abnormal hemoglobin for the red blood cells. The malaria parasite loves red blood cells, but, in people with sickle cell anemia, the malaria parasite has difficulty obtaining a foothold. Unfortunately, individuals with sickle cell anemia have severe health issues because they do not have normal hemoglobin. Nevertheless, people with sickle cell anemia are largely protected from malaria. A single variant gene provides, at a health related cost, protection from a debilitating parasite.

Baldness is much more complicated and clearly a multi-genetic trait. Therefore, let's step out of the box and think about why male pattern baldness occurs.

The "greater virility" hypothesis

NB: The section that you are about to read is a hypothesis from the author of this book. It may or may not have any validity, but please read it anyway.

Back when those hunter-gatherers were wandering west with their tribes, they probably did not go entirely bald. Most male hunters did not live until they were 50, 60, 70 or 80 years old. Men lived until they were 30 years old, if they were lucky and were not eaten by something that they were chasing. Any evolutionary selection for the characteristics of male pattern baldness must have occurred in young men, likely around the age of mating. At the time that they are seeking a mate, some men's hair shows the beginning of baldness, recession of the hairline at the temples.

Maybe a bowling ball head is not the point of the genetic changes that lead to male pattern baldness. In our attempts to understand why there is male pattern baldness, we may be looking in the wrong place on a man's body (his head) and at the wrong time in a man's life (older age).

Let's assume that the male baby, who will eventually go on to becoming bald, is born with the variant gene for the androgen receptor. Until puberty occurs, there is probably little consequence of carrying the variant gene. During puberty, androgens flood the bloodstream and affect many tissues to make the male a sexual machine.

Perhaps, in the old days, in the wandering tribe days, males born with the genetic variant of the gene for the androgen receptor, causing more androgen receptors to be made, had greater responses to circulating androgens. During puberty, these men would have had greater muscle mass, better bone structure, more sperm and increased sexual drive compared to men with the normal gene for the androgen receptor. Such men would be seen by the available females (and maybe some unavailable females) as better procreators, likely to have more children and more healthy children. And, suppose the slight recession of the hairline at the temples was the visible signs of this increased virility. Any woman would see it and go for it. As we discussed for hair of different colors, this is sexual selection driving evolution and would make a new trait become more prevalent in the population. Perhaps even in our times, men with receding hairlines are more sexy in their younger years. Think Cary Grant, Sean Connery, Burt Reynolds, Bruce Willis, Kevin Costner, John Malkovich, Alec Baldwin and Andre Agassi. Baldness may have little to do with old age and everything to do with coming of age.

Thus, one can easily postulate that male pattern baldness is what happens when you live too long after having shown the signs of increased virility at a younger age. This is an example of the phenomenon known as pleiotropism; a gene that was helpful at an early age becomes a problem-causing gene at an older age. Certain cancer causing genes are believed to be pleiotropic: early in life – useful; later in life – harmful. Perhaps pleiotropism is why many of us are experiencing balding as we age. So, guys, stop suffering over the loss of your hair. Suffer over the loss of the peak of your sexuality.

The "older is better" hypothesis

Another sexual selection hypothesis tries to explain baldness in older males as an enhanced signal of aging and social maturity, which is associated with decreased aggression and risk taking, and increased nurturing behaviors. In a psychosocial study, men and women viewed male models with different levels of facial hair (beard and moustache, or none) and cranial hair (full head of hair or receding hairline and bald). Participants rated each combination on 32 adjectives related to social perceptions. The combination of a beard and a full head of hair was seen as being more aggressive and less socially mature. Males with facial hair and those who were bald, or had receding hairlines, were rated as being older than those who were clean-shaven or had a full head of hair. Moreover, baldness alone was associated with more social maturity.

In our hunter-gatherer days, male pattern baldness may have conveyed the signal for an older male with greater social status, reduced physical threat and a willingness to raise offspring to adulthood. If most men 35,000 years ago did not live long enough to go bald, the one or two of them that did live into old age could have become the wise, bald elders of the tribe.

The "no prostate cancer" hypothesis
If you are a bald man, there may be reason to not worry so much about your prostate in the later years. Another hypothesis postulates that male pattern baldness evolved to protect men from prostate cancer. High levels of vitamin D are associated with a lower risk of prostate cancer. Because Vitamin D is activated in the skin by sunlight, a bald man with a large surface area of bare skin up top who stands, walks or works outside should have higher levels of vitamin D and a reduced risk of prostate cancer. A recent epidemiological study demonstrated a lower risk of prostate cancer in men with male pattern baldness. Of course, other factors may also be relevant.

WIDOW'S PEAK
Although not a receding hairline, a widow's peak is a downward "V" shape in the middle of the hairline. The term was originally used in the 1800's when this type of hairline was believed to indicate that a woman would outlive her husband. The shape is similar to a style of headdress worn by women in mourning. Despite the name, both men and women can have a widow's peak hairline. In pop culture, a widow's peak is often prominent on villains, like Count Dracula, Hannibel Lecter and The Joker. There has been debate as to whether a widow's peak hairline is determined by one gene. It is clearly inherited as a dominant trait, but more than one gene is likely to be responsible.

BETTER LIVING THROUGH CHEMISTRY
The pharmaceutical industry has come to the rescue of many who look in the mirror each day and suffer over their progressing baldness. Men with male pattern baldness may have low self-esteem, may feel depressed, may become introverted and may have feelings of unattractiveness

when they are caught in the frame of that never lying reflection. There were times when men covered up their hair loss, but now they have drugs.

Minoxidil (Rogaine, Regaine, Avacor and other trade names) is a serendipitous blockbuster. Although originally being developed for the treatment of high blood pressure, minoxidil had an unusual side effect during clinical trials. Completely unexpectedly, minoxidil grew hair on men's heads. Fortunately, insightful executives at the then pharmaceutical giant Upjohn took this side effect seriously, obtained a patent on the drug and its hair growing effects, formulated a topical (local) delivery vehicle (foam) and launched a proprietary, prescription medication for a medical condition that does no known physical harm. This success story is a powerful example of how the pharmaceutical industry can affect our lifestyle.

Minoxidil is believed to increase blood flow to the hair follicles in the scalp. Other mechanisms of action have been offered. Minoxidil maintains and strengthens remaining hair follicles, but does not cause new hair follicles to form or regrowth of new hair if the hair follicles are gone. How increased blood flow could stabilize hair follicles that are genetically programmed to miniaturize is not known. Interestingly, even after more than 20 years of use by the public, we do not completely understand how this drug works to make the big bucks. In 2006, the FDA approved minoxidil for over-the-counter use based on its strong safety record and ease of at home diagnosis (looking in the mirror).

Finasteride (Propecia) is a contender to minoxidil, but as of 2012 was only approved for oral use and not for topical use. Finasteride inhibits the enzyme, reductase, which converts testosterone into dihydrotestosterone, suggesting that dihydrotestosterone is an important driving force for androgenetic alopecia. Because of systemic side effects of oral finasteride (impotence, decreased libido, decreased ejaculate volume), men whose baldness might be helped by this drug choose to live with their bald heads rather than lose their sexuality. After all, what's the point? This drug will only become competitive with minoxidil when finasteride can be applied directly to the head.

Recently, a new pharmaceutical product has come on the market, Latisse, that can make eyelashes grow longer and thicker. The initial observation was quite serendipitous. The original product was an eyedrop that was used to treat glaucoma. One day, a woman with glaucoma in only one of her eyes, returned to see her ophthalmologist and asked if she could use the glaucoma eyedrop that he had prescribed in her other eye. When the doc asked why, she showed the ophthalmologist the luxuriant eyelashes that had grown in the eye receiving the glaucoma medication. The birth of a new cosmeceutical based on a side effect of a glaucoma drug. Women are rejoicing and men may soon be rubbing a son-of-Latisse on their balding heads.

BUYING NEW HAIR

A very popular alternative in the management of androgenetic alopecia is to buy new hair. This can be done as a transplant, wig or spray on hair cosmetic. Hair transplant surgery came of age in 1959 with the publication of the work of the founding father of this answer to baldness, Dr. Norman Orentreich. Dr. Orentreich demonstrated donor dominance, meaning that only hair follicles taken from an area of the head where hair was still growing would continue to grow when transplanted into an area where hair had stopped growing. To restate this point, hair follicles transplanted to the top of the head must come from regions of the head that are still growing hair. Hair transplants from parts of the body other than the head (chest, arm, leg) do not result in successful, aesthetically pleasing scalp hair.

Over the last 50 years, hair transplant surgery has been fine-tuned. Using microsurgery, follicular units are transplanted. A follicular unit contains 1-4 hair follicles with associated sebaceous glands, a vascular and neural plexus and a minimal amount of surrounding connective tissue. A highly, technically trained team harvests the donor graft material as an excised strip or as tiny plugs of tissue from the scalp on the back of the head. The strip is approximately 7 inches long by a half inch wide and contains 1100-1800 follicular units, which are then individually microdissected out as slivers. The sites in the recipient regions for transplant (the bald areas) are prepared by needle puncture. In the first session, about 1200 follicular units are transplanted to the prepared sites. Usually, the first step is to recreate the hairline and then the patient has 2-3 more transplant sessions separated by about one year.

OR, SHAVE IT OFF
At the beginning of the 21st century, many men who were going bald took the matter into their own hands. Even more than exercising their free-will, they thought: I am in control. They shaved their heads. No more thinning hair, office nerd-look for them. Men who changed the world, like Steve Jobs of Apple and Jeff Bezos of Amazon, rejected the horseshoe shaped ring of remaining hair around their head that would make them appear as a grandpa. Also, in popular American culture, the super-buzz cut portrays hypermasculinity, strength and confidence: Bruce Willis versus George Costanza. The new, bold, aggressive, don't-mess-with-me look has become a hairless dome encasing a seemingly outsized brain. These men's dominant heads project: I am the master of my fate.

PREGNANCY
The radiance of a pregnant woman is due, at least in part, to her flourishing hair. The head hair of pregnant women may be temporarily subject to systemic control. During pregnancy, there can be a hormone stimulated increase in the percentage of hair follicles of the head in the anagen phase of the hair growth cycle, perhaps 85% going to 98%, causing a pregnant woman's hair to appear more dense and thicker. However, the pregnancy-induced state of more working hair follicles does not persist. Following childbirth, many hair follicles on the head enter the telogen phase, which leads to loss of hair at about 3 months postpartum. Women claim that their hair is coming out in clumps. Eventually, the hair growth cycle readjusts itself from the flood of hormones and the number of growing hairs usually returns to pre-pregnancy levels. Occasionally, some women find that their

hair is thinner after a pregnancy. In these cases, there was likely a reduction in hair follicle density on the head during the second and third trimester. Thin hair after pregnancy will remain an ongoing condition because new hair follicles are not made to replace the ones that were lost.

CHEMOTHERAPY

Chemotherapy, endured by cancer patients, have significant effects on the hair follicles, and the hair loss that is associated with chemotherapy can be devastating to the psyche of the patient. Hair loss during chemotherapy negatively affects the patient's perception of their appearance, body image, sexuality and self-esteem. Patients feel deprived of their privacy because extensive hair loss is readily interpreted by the public as being associated with cancer. On rare occasions, women have declined chemotherapy because of the trauma of hair loss and other side effects.

The chemotherapeutic drugs stop the rapid division of cells in the hair follicle that is needed to continually grow the hair shaft longer. The anti-cancer drugs interfere with the hair growth cycle and the root of the once growing hair shaft becomes weak, constricted and thin. Subsequently, the hair shaft breaks, while in the hair follicle, and falls out. The loss of hair is during the anagen growth phase of the hair growth cycle, which is different than the normal shedding of hair during the telogen phase. Because 80-90% of scalp hair is in the anagen phase at any given time, the hair loss associated with chemotherapy is copious, usually beginning at 1-3 weeks after initiation of treatment, and is complete within 1-3 months. The hair loss becomes quite noticeable when about half of the hair is gone.

After cessation of chemotherapy, hair growth resumes in 3-6 months. The number of hair follicles is usually not altered by chemotherapy and the hair growth cycle returns to normal. However, the new hair that is growing in may have a changed color, structure, texture and/or density.

There is a treatment to prevent chemotherapy-induced loss of hair, although its use and effectiveness is controversial. Scalp cooling is done with either cooling agents applied with a cap or by continuous cooling of the scalp with cold air or liquids. The cooling treatment is an attempt to cause constriction of the scalp blood vessels when the chemotherapeutic agent is in the bloodstream, so as to limit the exposure of the hair follicles to the toxic substance. Although positive results have been reported in several large clinical studies, for some reason, perhaps financial, this auxiliary treatment is not generally available at cancer centers.

PULLING HAIR

Cutting hair is not painful (because hair is dead) but pulling hair is painful, in some people more that others. There are no real pain receptors in a hair shaft or in a hair follicle, but pulling on the hair shaft elevates the skin in which the hair follicle sits so that cutaneous pain receptors are activated. When pulling out a hair, a slight popping sound is heard as the hair shaft separates from the hair follicle, often accompanied with an involuntary: "Ouch!" Most people find hair pulling a momentary annoyance. Some people get off on it.

Trichotillomania is a psychological disorder in which the person obsessively and compulsively pulls out their hair, one at a time, to feel the pain. The uncontrollable desire to yank out one's hair is believed to affect perhaps 2 million American adults, mostly women, and probably even more children. Exact numbers are hard to come by because people with the condition often hide their embarrassing, patchy bald spots by wearing a hat. Stress and boredom can make the person with TTM pull more hairs. In a small number of cases, a variant gene is associated with this psychological disorder. Anti-depressants and other drugs have been tried but have not proven effective.

While we are on the subject, not that you have ever seen an S&M porn flick, but there is usually a scene in which a woman's hair is being pulled, enhancing her pleasure as she screams in rapturous agony. The woman is mounted doggy-style while the man grabs her hair in a thick bunch from the nape of the neck and pulls backward. Some women readily report that hair pulling heightens their orgasm and men describe experiencing increased arousal. The woman feels taken and the man feels dominant. Maybe this is part of a woman's fantasy, choosing to give up her free will. As in other S&M acts, a certain amount of peripheral pain, transmitted to the brain by nerves in the skin, increases the pleasure associated with sex. Whether the gender roles can be reversed, a woman pulling a man's hair to reach heightened orgasm, is not clearly answered in porn movies.

OTHER WAYS TO LOSE YOUR HAIR

Traction alopecia is most commonly found in people with ponytails or cornrows which pull on the hair with excessive force. In these conditions the hairs are not absent from the scalp but are broken. When the hairs break near the surface of the scalp, they become typical, short, exclamation mark hairs.

Certain fertility stimulating drugs can cause hair loss. Traumas such as major surgery, poisoning, and severe stress may cause a hair loss condition known as telogen effluvium. Iron deficiency is a common cause of thinning of the hair, though frank baldness is not usually seen. Some fungal infections can cause massive hair loss. Temporary loss of hair occurs in areas where sebaceous cysts persist for one to several weeks. Radiation to the scalp, as happens when radiotherapy is applied to the head for the treatment of certain cancers, may cause loss of hair in the irradiated areas.

Alopecia areata is an autoimmune disorder also known as "spot baldness" that can result in hair loss ranging from just one location to every hair on the entire body. Localized or diffuse hair loss may occur in cicatricial alopecia, lupus erythematosus, lichen plano pilaris, folliculitis decalvans or postmenopausal frontal fibrosing alopecia. Tumours and skin outgrowths (basal cell carcinoma, squamous cell carcinoma, sebaceous nevus) induce localized baldness. Hypothyroidism can cause hair loss, typically frontal, and is peculiarly also associated with thinning of the outer third of the eyebrows. Hyperthyroidism can cause hair loss on the top of the head.

EPIDEMIOLOGICAL ASSOCIATIONS

Epidemiologists love to make correlations. Androgenetic alopecia has been associated with coronary heart disease in men and women, insulin resistance, diabetes and the related obesity and lipid disorders, high blood pressure, and benign prostatic hyperplasia. None of these studies show that any of these diseases actually causes androgenetic alopecia. They are just epidemiological associations. What anyone can make of these correlations and associations remains to be seen.

Smoking may be a factor associated with age-related hair loss. A study of Asian men, which controlled for age and family history, found statistically significant, positive associations between moder-ate or severe thinning of hair and amount of smoking.

Smoking may be a factor associated with age-related hair loss. A study of Asian men which controlled for age and family history found statistically significant, positive associations between moderate or severe thinning of hair and amount of smoking.

BALDNESS FOLKLORE

There are many myths regarding the possible causes of baldness and its relationship with one's virility, intelligence, ethnicity, job, social class or wealth. While skepticism is warranted due to lack of scientific validation, some of these myths may have a degree of underlying truth.

"You inherit baldness from your mother's father." As discussed earlier in this chapter, there is a genetic basis, in some grandsons, for this belief. However, the statement is not absolute. In many families, there are bald maternal grandfathers who have grandsons with a full head of hair. Several gene variants need to be inherited from both mother and father to produce a son who will grow up to have male pattern baldness.

"Intellectual activity can cause baldness." In the classical world, it was thought that if a person was bald it was likely that he had an adequate amount of fat in his diet. High fat in the diet meant that mental development was probably not stunted by malnutrition during the crucial formative years. In addition, this person was more likely to be wealthy and to have had access to a formal education. Furthermore, social standing, accrued through intelligence, might have compensated for decreased physical attractiveness associated with hair loss. The combination of baldness, intellect, wealth and social status would have produced male offspring who were prone to both high intellect and hair loss. Thus, the bald man in ancient Rome was probably smart. However, lifestyle is less likely to correlate with intelligence in the modern world.

"Baldness can be caused by not enough sex, sexual frustration or frequent ejaculation." A low sex drive and sexual frustration associated with decreased performance may be due to decreased circulating testosterone. Depending on frequency, ejaculation can raise or lower blood androgen levels. But, as discussed earlier, circulating androgens are not what cause balding. Admittedly, these are very difficult human situations to examine by research studies. By the way, frequent masturbation does not cause the hair on the back of your hands to grow.

"Bald men are more virile or more sexually active than others." Levels of testosterone are strongly linked to libido and there are decreased circulating levels of androgens as we age. As discussed above, virility and sexual activity may be higher in men in their 20's who will go on in life to develop male pattern baldness. However, sexual activity is multi-factorial and not only androgen dependent. Because hair loss is progressive, a person's hairline may be more indicative of their past performance than their present disposition.

"Tight hats cause baldness." This may not be such a myth as hats do cause hair breakage and, to a lesser degree, split ends. Because hats are not washed as frequently as other clothing, wearing hats can lead to scalp uncleanliness and possible parasite contamination in people with

naturally oily scalps. Some scalp infections, if left untreated, can cause hair loss.

All those bald men that we see everyday don't really have much of a choice about hairstyle. The horseshoe shaped head of remaining hair is in their genes, even though we are not exactly sure of why and how. But all men, even bald men, have a choice about how to "wear" their facial hair. Let's move on to moustaches and beards.

8

FACIAL HAIR (HIS AND HERS)

"Keep the beards and cut the moustaches short."

-the Prophet Muhammed

What can be generally considered facial hairs are, in descending order, eyebrows, eyelashes, moustache and beard. Nose hair and ear hair are hard to classify. By the definition used throughout this book, only the moustache and beard are truly hair because they continue to grow. Yes, of course, there are professors with bushy eyebrows, but these hairs do not grow very long. Women may curl their eyelashes but they certainly never cut them.

MOUSTACHES AND BEARDS

Facial hair, the moustache and the beard, are a male thing, although both men and women have vellus hair on their face. Not counting the approximately 110,000 hair follicles in the scalp, there are about 800,000 hair follicles covering the rest of the head, most of them producing vellus hair. In many places on the face, the vellus hair is not visible to the naked eye. Women go through puberty and, for most of them, nothing much happens to those vellus hairs. Men go through puberty and the vellus producing facial hair follicles turn into hair follicles that make the thick, terminal hairs of the beard. Facial hair in the mature male grows at about the same rate as hair in the scalp (1/2 inch per month). Some men have very dense, thick moustaches and beards, and some men have very few hairs in their moustache and beard areas.

How much terminal hair a man will grow under his nose and on his cheeks, jaw and neck is related to the geographic area of origin of a man's ancestors and his particular complement of hair-related genes. Men of European ancestry tend to have thick moustaches and beards although there are many exceptions. Men with origins in the Middle East and the Indian sub-continent generally have thick moustaches and beards; whereas most men of Asian ancestry have thin, wispy moustaches and chin hair but not much in the way of full beards. Men of African origin have thin, short moustaches and sparse beards with similar curliness to the hair on their heads.

The longest moustache in the world drapes from the upper lip of Badamsinh Juwansinh Gurjar, a man from Ahmedabad, India. In 2004, his moustache measured 12.5 feet on each side. He had not cut his moustache in 22 years. A Norwegian-American, Hans Langseth, holds the record for the longest beard at 18.5 feet. When he died in1927, his family cut off his beard, all but twelve inches of it. The beard can be viewed in a 1967 photograph of a group of physical anthropologists holding 17.5 feet of it at the Smithsonian Institution.

In common speech, a man can *have* a moustache or a beard, *wear* a beard, or *sport* a moustache. Unusual usages. Whiskers usually refer to the hair on a man's chin or cheeks. The use of the term beard is a bit indefinite. Clean-shaven men often refer to their thick beards that need a shave. A man with both a moustache and beard is referred to as having a beard. However, Abraham Lincoln had a beard without a moustache, a popular style of the time. At least when a man sports a moustache, it is clear that he only has a moustache.

FACIAL HAIR (HIS AND HERS)

The beard that a man grows has color. The same types of pigments that bring natural color to scalp hair also provide natural color to beard hair. In men of European ancestry, sometimes the scalp hair color and the beard hair color do not match. A man with dark brown hair may grow a beard that has predominantly light brown hairs, with some reds and even blonde hairs mixed in. A man with red hair may grow a beard with red hairs of a different hue than the hair on top of his head. A man with blonde hair may have a beard of a darker color. Why there is often a mismatch between scalp hair and beard hair in men of European origin is not known. When gray hairs fill in a man's beard, his face looks aged. There are a few products that claim to color beard hair. However, gray beard hair grows almost three times faster than non-gray beard hair and, like gray hair up top, is difficult to color. Most men who try to add color to their beard report unsatisfactory results.

Some men grow beards because they do not like to shave. Many more men grow beards because they like the way they look in a beard. Thus, the free will act of growing a beard is laden with vanity. And, there is much maintenance. The long flowing beards of yesteryear have given way to trimming the beard into specific shapes. Even the 5 o'clock shadow or "two day growth," which is particularly popular with today's young men, needs upkeep. But for sheer maintenance, such as the need for specialized wax, nets, brushes, combs and scissors, nothing requires as much attention on a man's face as a lovingly, carefully and personally shaped moustache. It must be narcissism. What else can account for the many styles, sometimes outrageous, but so blatantly visible above the upper lip?

EVERY WOMAN HAS HER FAVORITE TWEEZERS

Most of this chapter is about men. Women typically do not have terminal hair on their face, but have vellus hair on their face and covering most of their body. Nevertheless, women are capable of developing facial hair. In certain ethnic groups, thin, dark, facial hair is noticeable in young women above the upper lip and on the sides of the face. This normal hair growth does not get very long and can be removed, but generally is not trimmed. Excessive facial hairiness in women is sometimes an indication of pathological hormonal variations. For some reason, Shakespeare imagined the witches in *Macbeth* to have hairy faces, as Banquo says: "You should be women, and yet your beards forbid me to interpret that you are so."

For many women, especially after menopause, aberrant hairs can grow almost anyplace on the face, though typically significantly less than what grows on men. It is likely that the drop in estrogen associated with menopause permits the hair follicles on the face of women to respond to the low levels of circulating androgens. Hair on the face of a woman is unwanted and is usually removed by tweezing, depilatories or waxing. The point is to get rid of it in most cultures; it can be a considerable social stigma. The vainglorious facial hairs on a man are not usually celebrated on a woman's face.

BEARDED WOMEN
Excessive androgens are thought to be the underlying malady suffered by bearded women (hypertrichosis). Circuses once displayed bearded women as freaks. The most famous bearded woman was Annie Jones who was part of PT Barnum's "side show." Annie had bushy sideburns, a long flowing moustache and a raggedy beard about 8 inches long. She claimed to have turned down painters who wanted her to model for portraits of Jesus. Annie was married twice. After 15 years of marriage to Richard Eliot, she divorced him for her childhood sweetheart William Donovan. Obviously, Annie had other, more subtle attractive qualities.

ANDROGENS, PUBERTY AND FACIAL HAIR
Primary sex characteristics in a human male are the equipment that he is born with: a penis and testicles. Facial hair is a secondary sex

characteristic in human males because it develops after birth at the time of puberty, when the male equipment becomes ready to act. Adolescent males usually start developing facial hair in the later years of puberty between 15-18 years of age, and most do not finish developing a fully adult beard until their early 20's or even older. However, some boys may develop facial hair as young as 10-12 years old. The facial hair that develops during puberty can be thin, sparse, bushy or bristly.

On the face of a young male, the region where hair appears first during puberty varies among individuals. Most often the moustache is formed before the beard and the process is as follows: The first facial hair to appear tends to grow at the corners of the upper lip and then spreads to form a moustache over the entire upper lip. This is followed by the appearance of hair on the upper part of the cheeks and the area under the lower lip. Sideburns get longer. Eventually the beard hairs spread to the sides, the lower border of the chin and the lower face, and a full beard is formed. During this process, the ill at ease, lanky teenager is left to wonder: When do I start shaving?

The beard develops during puberty, which means that the hair follicles on the face are sensitive to androgens. Therefore, the hair follicles on the face have a different sensitivity compared to the hair follicles on the scalp, which are not sensitive to androgens during puberty. Furthermore, the facial hair follicles respond to androgens circulating in the bloodstream, again unlike the hair follicles in the scalp of balding men that putatively respond to locally made androgens. Thus, beard growth is linked to stimulation of hair follicles in the face by increased blood levels of dihydrotestosterone at the time of puberty. The levels of dihydrotestosterone can vary with sexual activity, aging and even with the season. With age, there is less beard growth and there is some thought that beards grow faster in the summer months. How fast the beard grows is also genetic. If you are a man and your father has a heavy, thick fast growing beard, chances are you will too. In many young men of European ancestry, the rate of facial hair growth may actually be a little bit faster than the rate of scalp hair growth.

As far as anyone can tell (and not many people have done careful biological studies), the hair follicles that produce the hairs of the beard are the same in microscopic structure and function as the hair follicles in the scalp that make the head hairs. However, the shape of beard hairs is often different than scalp hairs. Moustache and beard hairs are generally oval in cross sectional shape and the openings of the hair follicles are oval. There is obviously a long growth phase (anagen) for beard hairs because they will grow to the waist and even lower. How long of a resting phase (telogen) occurs after the growth phase of beard hairs has not been studied.

IF IT WASN'T FOR THE ANCIENT GREEKS

The Greek word for beard is *pogon* which is the root of a number of words relating to beards that are not heard in every day conversation. For example, the study of beards is called "pogonology," and one who studies beards is a "pogonologist," of whom there are not many. Most pogonologists work for large companies that produce hair care products and equipment. Extending the use of the Greek root, "pogonotomy" is the technical term for shaving and, conversely, "pogonotrophy" is the technical term for growing a beard.

SHAVING

For most men, pogonotomy is a ritualistic behavior. We can consider shaving a ritual because it is a frequent behavior performed in a fixed, sequential series of steps with a specific group of tools and chemicals whose sole purpose is to accomplish the achievement of being clean-shaven, and thereby bringing the ritualistic behavior to a satisfying conclusion. Interruptions or deviations in the series of steps are not welcome. Changes in the series of steps or changes in the ritualistic tools and chemicals are only made after due consideration. Each man's shaving ritual is slightly different than another man's shaving ritual. Variations can include time of day, bodily coverings, choice of mirrors,

FACIAL HAIR (HIS AND HERS)

before/during/after bathing or showering, pre-shave facial treatments, creams, lathering, shaving utensils, post-shave ablutions, and after shave treatments and fragrances. Nevertheless, once a man's individual ritualistic behavior is established, it is adhered to almost religiously.

The purpose of shaving is to cut the growing hair shaft as close as possible to where it is just emerging from the mouth of the hair follicle in the skin. There are so many variations on how to do this that to try to describe them in this book would be endlessly boring. If you are male, you need no further description. If you are a curious or uninformed female, go watch your husband, significant other, brother, father, son, cousin, uncle, friend, business associate, girl friend's husband or any other male in your life. Or, find and watch the once common but now few barbers that are still giving shaves to men. If you really have nothing to do, be an anthropologist and go watch them all, and take notes on the differences in the rituals.

At least some prehistoric men shaved. Artifacts from very early cultures include shaving blades that are sharpened seashells, flint, copper, bronze, iron or volcanic glass. Scraping and tweezing, for example by two seashells held together, was what cavemen endured to remove hair from their face. Not pleasant.

We know that the leaders of the Roman Empire were shaved by barbers who did not have steel for making razors, so they used bronze or copper blades. Some Roman barbers had reputations as clumsy butchers who left their patrons scarred about the cheeks and chin. Apparently, Julius Caesar preferred plucking out the hairs of his beard with a tweezers. Depilatories were also offered for those customers who refused the razor. In ancient Rome, the first occasion of shaving was regarded as the beginning of manhood, and the day on which the first shave took place was celebrated as a festival honoring the young Roman. Usually, this was done when he assumed the toga of a man.

Steel, which became available in the Middle Ages, allowed the honing of a sharp edge that would not readily become dull. The use of steel led to the development of the long beautiful swords of yore and the straight edge razor. The sharp steel razors in the hands of the barbers gave them the tools and, therefore, the license to be the local surgeons. Here is a bizarre amalgam of hair and steel. Remember that redheads are special. It was generally believed among metallurgists in medieval times that the best steel was made by cooling the glowing hot blades in the urine of redheaded boys.

In 1847, William Henson came up with the razor design we know today, the blade attached crosswise to a handle that resembles a small garden hoe. Shortly after, the Kampfe Brothers developed the guard along the blade called the safety razor. In 1902, King Camp Gillette (yes, his first name was King) invented the disposable razor and established a new marketing ploy: make money on the refill. Apparently, Mr. Gillette, a salesman and inventor at the time, first had the marketing and sales idea of making money on a replaceable part, but he did not know what to sell. One day, his razor went dull and a light bulb went off in his head: replaceable blades. An MIT engineer helped design the equipment that could mass produce the razors with disposable blades. During WW I, the US government ordered 3.5 million razors and 36 million blades from Mr. Gillette's new company. Today, the Gillette Company is the largest manufacturer of shaving products in the world, and is a great example of American ingenuity and marketing acumen.

Jacob Schick used his familiarity with repeating rifles in the military to invent the cartridge or injector razor but his real interest was in developing a dry razor. The first electric, dry razor, a head with rapidly moving blades attached by a flexible wire cable to an electric motor, was introduced in 1927. The response of the shaving public was underwhelming. Further improvements finally made the electric shaver a competitive product. In 1935, Mr. Schick moved to Canada (and his money to the Bahamas) to avoid US prosecution for tax evasion. The Schick Shaving Company made it big on electric shavers and disposable razors, in spite of Mr. Schick's absence, and today seems to follow closely the Gillette Company's innovations in razors.

Over the years, innovative inventors have changed the way men shave and the products they use. Today, if you are using a disposable razor, it is likely that you are holding a razor with five stainless steel blades lined up in a row. The Romans would have thought you a god.

Many of us have always believed that the more we shave our beard, the thicker the beard will become. But, the number of hair follicles cannot increase and the facial hairs themselves do not become thicker with frequent shaving. After facial hairs have grown for a day or two, the ends become tapered and they can feel softer. When shaved with a fresh blade, the newly growing beard hairs have cut, not yet tapered ends that appear thicker, and feel coarser due to the sharper, unworn edges. Shorter facial hairs are harder and less flexible than longer beard hairs, and look darker because they have not been bleached by sun exposure. So, beards do not get thicker when you shave them, it just feels and appears that way. Remember that scalp hair does not become thicker or fuller after shaving the head either.

For the same reasons, if a man wears a beard, the hairs of the beard will feel softer before he trims, not afterwards. The scissor or trimmer bluntly transects the long facial hairs leaving the ends un-tapered and sharp feeling.

MOUSTACHES AND BEARDS IN HISTORY

The history of moustaches and beards is all about the changing ideals of male vanity. In the course of history, men with facial hair have been ascribed various attributes such as wisdom, knowledge, sexual virility, masculinity, high social status and/or wealth. Conversely, men with facial hair have also been described throughout the ages as filthy, crude or having an eccentric disposition.

The ancient world

In ancient Egypt, a metal false beard was a sign of sovereignty and was worn by royalty, and sometimes cows. This *postiche* was held in place by a ribbon tied over the head and attached to a gold chinstrap. Look for it on the chins of pharaohs next time you go to a museum.

From about 3000 to 1500 BCE, the highest ranking leaders of western civilizations grew very long beards on their chins which were often dyed or hennaed, and sometimes woven with gold thread into plaits. The Persians were fond of long beards. Mesopotamian civilizations (Sumerian, Assyrians, Babylonians, Chaldeans and Medians) devoted great care to oiling and dressing their beards, using tongs and curling irons to create elaborate ringlets and tiered patterns.

To the classical Greeks, the beard was a sign of virility and a smooth face was regarded effeminate. The Spartans punished cowards by shaving off a portion of their beards. From the earliest times, a beard without a moustache was the common style and Greeks frequently curled their beards with tongs. In Homer's writings, a form of greeting was to touch the beard of the person addressed. In ancient Greece, an untrimmed beard was a sign of mourning.

Alexander the Great introduced the smooth shaven face to western civilization. He himself did not wear a beard. Historians tell us that Alexander ordered his soldiers to be clean-shaven, fearing that their beards would serve as handles for their enemies to grab and to hold the soldier during battle. The practice of shaving spread from the Macedonians, whose kings are represented on coins with smooth faces, throughout the known world of the empire. Aristotle conformed to the new custom, unlike other philosophers who retained the beard as a badge of their profession.

During the Roman Empire, beards were in and out of fashion. Beards were common amongst the Romans during their early history until being clean-shaven became a mark of the elite at the height of the Roman Republic. Being clean-shaven was a sign of being Roman and not classical Greek, particularly important as at that time the Romans thought the earlier Greeks were barbarians in many of their behaviors. In Rome, a long beard was considered a mark of slovenliness and squalor, but at times of mourning, beards were allowed to grow.

In the second century CE the Emperor Hadrian, was the first of all the Caesars to grow a beard, perhaps to hide scars on his face. Rome entered a period of widespread imitation of classical Greek culture, and many men grew beards. Until the time of Constantine the Great, the emperors appear in busts and coins with beards, but Constantine and his successors at the end of the 6th century were beardless.

To the east, in ancient India, the beard was allowed to grow long as a symbol of dignity and of wisdom. Beards were treated with great care and veneration, and the punishment for lacking moral restraint or committing adultery was to have the beard of the offending man publicly cut off. There was such a high regard for the preservation of beards that a man might pledge it as collateral for the payment of a debt.

Further to the east, moustaches and goatees were sprouting. Many Asian men cannot grow the thick beards of European men, but their moustaches and chin hairs drooped to great lengths. In 200 BCE, all of the thousands of terra cotta soldiers created for the first Emperor of China, Qin Shi Huang, and then buried underground in what is today Xian, sported handsome, flowing moustaches.

Two thousand years ago in the lands that became known as North, Central and South America, the indigenous people had little to no facial hair. There are no signs of moustaches or beards in the artifacts left behind by the Olmecs, Mayans, Incas, Aztecs or various other civilizations. Apparently, the tribes migrating across the Bering Straits to originally populate the New World came from regions of Asia where facial hair on men was minimal. But virility? Not a problem.

Celts and Germanic tribes

In old Greek sculptures of Celts, they are portrayed with long hair and moustaches but without beards. However, among the Celts of Scotland and Ireland men typically let their facial hair grow into a full beard, and it was considered dishonorable for a Gaelic man to have no facial hair. In some Germanic tribes, a young man was not allowed to shave or cut his hair until he had slain an enemy. The Germanic tribe, the

Lombards, derived their fame from the great length of their beards (from *longobards* or *langbärte*, meaning long beard). Otto the Great swore by his long beard that covered his breast when he wanted his proclamations to be taken seriously.

Middle Ages

Charles IV, the Holy Roman Emperor, had a full beard, and kings and noblemen all had full beards. In medieval Europe, a beard displayed a knight's virility and honor. The Castilian epic knight El Cid is described as "the one with the flowery beard" and the chivalrous Don Quixote is always identified by his moustache and goatee. Grabbing somebody else's beard was such a serious offense in the Middle Ages that a duel might follow.

From the Renaissance to today

In the 15th century, most European men were clean-shaven; however, 16th century beards were allowed to grow to extraordinary lengths. Later, some beards began to be shaped: the Spanish spade beard, the English square cut beard, the forked beard and the stiletto beard. Shaped beards were especially prevalent during Queen Mary's reign. In urban circles of Western Europe and the Americas, beards were out of fashion after the early 17th century. In 1698, Peter the Great of Russia ordered men to shave off their beards and in 1705 he levied a tax on beards in order to bring the look of Russian society more in line with contemporary Western Europe.

The fashion of wearing a beard then declined in western society and in the early 18th century most men, particularly amongst the nobility and upper class, were clean-shaven. There was, however, a dramatic shift in the popularity of the beard during the 1850s. Beards were grown by many leaders, such as Alexander III of Russia, Napoleon III of France, Frederick III of Germany, as well as many leading statesmen and cultural figures, such as Benjamin Disraeli, Charles Dickens,

FACIAL HAIR (HIS AND HERS)

Giuseppe Garibaldi, Karl Marx and Giuseppe Verdi. The stereotypical Victorian male is the stern figure clothed in black whose gravitas is added to by a heavy beard. The same trend was seen in the United States in the post-Civil War presidents. Before Abraham Lincoln, no President had a beard. Everyone knows the story of the school girl who wrote to President Lincoln urging him to grow a beard to hide his less than handsome visage. For the next 50 years after Lincoln, every President except Andrew Johnson and William McKinley had either a beard or a moustache. Around the time of the Civil War, the goatee became popular in the US.

By the early 20th century, beards began a slow decline in popularity. Although retained by some prominent figures who were young men in the Victorian period (Sigmund Freud), most men who wore facial hair during the 1920's and 1930's limited themselves to a moustache or a goatee (Marcel Proust, Albert Einstein, Vladimir Lenin, Leon Trotsky, Adolf Hitler, Joseph Stalin). In America, meanwhile, popular movies portrayed heroes with clean-shaven faces. Concurrently, the mass marketing of products such as the Gillette Safety Razor, shaving creams and other aids was becoming pervasive. Short hair and clean-shaven faces were the only acceptable style for decades to come. The few men who wore a beard or portions thereof during this period were either old, central Europeans, members of a religious sect or in academia.

The beard was reintroduced into contemporary, mainstream society by the counterculture, first with the beatniks in the 1950's, and then with the hippie movement of the mid 1960's. In the mid to late 1960's and throughout the 1970's, beards were worn by hippies and businessmen alike. Following the Vietnam War, beards became very in. Popular rock, soul and folk musicians like The Beatles, Barry White and Peter, Paul (not Mary) wore full or partial beards. In American culture, the trendy beard faded in the mid 1980's.

From the 1990's onward, the fashion in beards has generally trended toward either a goatee, Van Dyke, or a closely cropped full beard undercut on the throat. By the end of the 20th century, the closely trimmed

Verdi beard, often with a matching integrated moustache, became relatively common. By 2010, the fashionable length for young hunks approached a two day growth.

MOUSTACHES AND BEARDS IN RELIGION

Ancient religions, unkempt beards

In the Old Testament, anyone who was anybody had a beard. God (according to Michelangelo), Abraham and Moses had full, long, unkempt beards. Interestingly, Adam is never depicted with a beard. In Greek mythology, Zeus and Poseidon are always portrayed with raggedy beards, but not Apollo. A bearded Hermes was replaced with the more familiar beardless youth in the 5th century BCE. Whether having a beard is associated with the spiritual and religious beliefs of these early patriarchs and gods, or was just the contemporary style because of not shaving, is untold in the lore of the times.

Judaism

Many Orthodox Jews grow facial hair. The Old Testament states in Leviticus that: "You shall not round off the side growth of your heads nor harm the edges of your beard." *Peyose*, long curls on the side of the head, are worn and beards are longer on the sides than in the center. The interpreters of the Talmud further explain that a man may not shave his face with a razor that has a single blade, because the cutting action of the blade mars the skin. Because scissors have two blades, some opinion leaders in Jewish law permit their use to trim the beard, as the cutting action comes from contact of the two blades and not the blade against the skin. For the same reason, Conservative Jews may use electric shavers to remain clean-shaven, as such shavers cut by trapping the hair between the blades and the metal grating, a scissor-like action. As in all Talmudic laws and interpretations, arguments about whether shaving the face or trimming beard hair is permitted or not, and if so, what kind of instruments can be used, have been going on for millennia. Certainly, no electric shavers on the Sabbath.

The Zohar, one of the primary sources of Kabbalah, or Jewish mysticism, attributes holiness to the beard, describing that hairs of the beard symbolize channels of subconscious holy energy that flows from above to the human soul. Therefore, most Hasidic Jews, for whom Kabbalah plays an important role in their religious practice, traditionally do not shave or even trim their beards.

Christianity
Jesus is almost always portrayed with a beard in iconography and art dating from the 4th century onward. Being Jewish, it is more than likely that Jesus really did have a full beard. The New Testament disciples of Jesus such as Saint Peter and John the Baptist had beards. John the Apostle is generally depicted as clean-shaven in Western European art to emphasize his relative youth. Eight of the figures in *The Last Supper* by Leonardo da Vinci are bearded. In Isaiah's prophecy of Christ's crucifixion, he describes Christ having his beard plucked by his tormentors.

Nowadays, beards are rare in the Catholic Church, although Franciscans wear a beard as a sign of their sect. At various times in its early history, the Catholic Church permitted beards but more recently has prohibited facial hair. In Eastern Christianity, beards were often worn by members of the priesthood and by monks. At times, beards have been recommended for all believers. Clerics in the Greek Orthodox Church always had long, full beards.

Among Protestants, growing a beard has been described as "a habit most natural, scriptural, manly and beneficial." Martin Luther provided the precedent, and then virtually all the continental reformers deliberately grew beards as a mark of their rejection of the old church. Clerics with beards were an aggressive anti-Catholic gesture as Protestant movements took hold.

Amish men shave until they are married, then grow a particular form of a beard that they maintain for the rest of their life. The full, rounded beard without a moustache is a symbol of their religious identity. In Ohio, the domineering leader of a renegade Amish sect was convicted under the federal hate-crimes laws after his followers invaded the homes of rivals, physically attacked them and removed their facial hair with shears used on animals. US Department of Justice prosecutors proclaimed that these were acts of terror and humiliation against people who just wanted to practice their religion and live in peace. The defense argued that the federal government should stay out of this local dispute, but a jury concluded that the cult members had gone a hair too far.

Rastafari is a Christian movement and ideology with a strong Afrocentric culture. A Rastafarian's beard is a sign of his pact with God. Not cutting head or facial hair is strictly interpreted from the Bible: "They shall not make any baldness on their heads, nor shave off the edges of their beards, nor make any cuts in their flesh." Rastafarians reject the comb, brush and scissors as tools of what they see as the highly corrupt western society. The curly hair, hanging in long, massive locks, is known as dreadlocks. The length of dreadlocks is a measure of wisdom, maturity, and knowledge in that it can indicate not only the Rasta's age, but also their time as a follower of the movement.

On the other end of the facial hair spectrum, Mormon men are strongly encouraged to be clean-shaven and formal prohibitions against facial hair are given to young men entering their two years missionary service. Students and staff of the church-sponsored Brigham Young University are asked to adhere to the Church Educational System Honor Code, which states: "Men are expected to be clean-shaven; beards are not acceptable."

Hinduism

For Hindus, the following of ancient texts regarding beards varies according to whom the devotee worships. Most original idols lack moustaches and beards. However, many Yogi practitioners grow beards and generally Sadhu ascetics have beards, as they are not permitted to own anything, which would include a razor, when they go into their cave or forest.

Islam
The Prophet Muhammad directly commanded his followers to grow ungroomed beards but to maintain their moustaches short. Muslim scholars view keeping a beard as being commendable for men, as it follows the example of Muhammad, and most consider it obligatory. In Islamic lore, God commanded Abraham to keep his beard, shorten his moustache, clip his nails, shave the hair around his genitals and pluck his underarm hair. We do not know for sure whether Abraham followed these commands.

Sikhism
Sikhism, a religion with very strict beliefs and practices, was started in India in the 1600's as a protest against Islam. In 1699, Guru Gobind Singh, the tenth and last Sikh Guru, ordained and established the keeping of unshorn hair as part of the identity and one of the signs of being a Sikh. Sikhs consider the beard to be part of the nobility and dignity of their manhood, and respect for the God given form of their bodies. *Kesh*, uncut head and beard hair, is one of the Five Ks, five compulsory articles of faith for a baptized Sikh. As such, a Sikh man is easily identified by his turban and untrimmed beard.

TODAY'S PUBLIC BEARDS
The US military and most police forces ban beards on the bases of both hygiene, and of the necessity of a good seal for gas masks and other headgear. In the military in countries throughout the world, beards are generally banned, but there have been exceptions especially in British troops in the last century when beards became an integral part of the uniform. Professional airline pilots are required to be clean-shaven

to facilitate a tight seal with auxiliary oxygen masks. Similarly, fire fighters may also be prohibited from growing full beards to obtain a proper seal with SCBA equipment. In sports, moustaches and beards may be banned or simply not be practical, except in baseball where they have become part of a baseball player's image.

One stratum of American and many other societies where facial hair is virtually nonexistent today is in government and politics. Would JFK have been elected if he had a moustache or beard? The last President of the United States to wear any type of facial hair was a century ago; William Howard Taft, who, with his flowing, English-style moustache, was in office from 1909 to 1913. The last Vice President of the United States to wear any facial hair was Charles Curtis, who, with his brush, chevron-style moustache was in office from 1929 to 1933. In the US today, we expect our politicians to have that clean-cut, all-American look.

MOUSTACHE STYLES

Some self-amusements on the upper lip:

- Mexican – Big and bushy, beginning from the middle of the upper lip and pulled to the side, the hairs growing slightly up beyond the end of the upper lip.
- Dalí – Narrow, long points bent or curved steeply upward; areas past the corner of the mouth are shaved. Artificial styling aids are needed. Obviously named after Salvador Dalí and perhaps unique.
- English – Beginning at the middle of the upper lip the whiskers are very long and pulled to the side, slightly curled. The ends may be pointed slightly upward and areas past the corner of the mouth are usually shaved.
- Imperial – Whiskers growing from both the upper lip and cheeks, curled upward, formerly seen on Russian tsars.
- Fu Manchu – Long, downward pointing ends, generally beyond the chin.

Pancho Villa – Similar to the Fu Manchu but thicker. A droopy moustache, generally much more so than that worn by the historical figure.

Handlebar – Bushy, with small upward pointing ends. Also known as a spaghetti moustache, because of its stereotypical association with Italian men.

Horseshoe – Popularized by modern cowboys and consisting of a full moustache with vertical extensions from the corners of the lips down to the jaw line, resembling a horseshoe. Also known as a biker's moustache.

Pencil – Narrow, straight and thin as if drawn on by a pencil. Closely clipped, outlining the upper lip, often with a shaven gap between the nose and moustache, and a gap between left and right sides of the moustache hairs.

Chevron – Thick and wide, usually covering the top of the upper lip.

Toothbrush – Thick, but trimmed to just above the upper lip.

Walrus – Bushy, hanging down over the lips, often entirely covering the mouth.

BEARD STYLES

To name a few:

Circle – A beard in which the moustache and beard below the mouth and chin are connected, but all other facial hair is shaved off.

Full – A downward flowing beard with either styled or integrated moustache.

Sideburns – Hair grown from the temples down the cheeks toward the jaw line. Sometimes with a moustache.

Chinstrap – A beard with long, thin sideburns that comes forward and ends under the chin.

Donegal – Similar to the chinstrap beard but covers the entire chin.

Garibaldi – A wide, full beard with rounded bottom and integrated moustache.

Goatee – A tuft of hair grown on the chin, sometimes resembling that of a billy goat.

Junco – A goatee which extends upward and connects to the corners of the mouth.

Hollywoodian – A beard with integrated moustache that is worn on the lower part of the chin and jaw area, without connecting sideburns.

Reed – A beard with integrated moustache that is worn on the lower part of the chin and jaw area, and tapers towards the ears without connecting sideburns.

Royale (Imperial) – A narrow pointed beard extending from the chin.

Stubble – A very short beard of only one to a few days growth.

Van Dyke – A goatee accompanied by a moustache.

Verdi – A short beard with rounded bottom and slightly shaven cheeks with prominent moustache

Neck beard (Neard) – The chin and jaw line shaven, leaving hair to grow only on the neck.

Soul patch – A small beard just below the lower lip and above the chin.

Mutton chops – Long, lamb chop shaped sideburns connected to a moustache, but with a shaved chin.

Stash burns – Sideburns that drop thinly down the jaw but jut upwards across the moustache, leaving the chin exposed.

Fortunately for men, beards do not require much care and upkeep, aside from a scissor or electric trimmer/shaver to occasionally cut and shape the beard, and a shampoo after eating an ice cream cone or a plate of spaghetti in tomato sauce. What men and women do to care for the hair on their scalp is another story or, in the context of this book, another (the next) chapter.

9

HAIR CARE, TRICHOLOGY, FUNGI AND NITS

"When I want to make front page news, I change my hairstyle."

- Hillary Rodham Clinton

Your hair requires a lot of attention, thought and work. You probably clean your house once a week (usually), wash your clothes every 7-10 days (maybe), fuss with your indoor plants every few days (enjoyable) and pay your bills monthly (necessary). But your hair is a daily chore that, for some, may take an hour or more each day. After all, your hair is your crowning glory.

Hair is the only body structure that is completely renewable by regrowth. Because of the hair growth cycle, hair can be subjected to insults in the name of treatments that could not be sustained by any other body structure. Constant renewal also means that any alterations in hair length, contour, color or texture are temporary until the altered hair is sloughed, cut or weathered.

WEATHERING

Hair damage results from thermal, mechanical and chemical injuries that alter the physical structure of the hair shaft. This damage is, for the most part, self-inflicted. For example, every time you wash your hair you are damaging your hair. If your hair is 12 inches long (shoulder length) and you shampoo every day, the outer few inches of your hair have been lathered in soapy solutions about 700 times over a two years period of growth. Imagine what a wool sweater would look like that had been washed 700 times. The hair care industry has coined the term "weathering" to refer to the progressive, degenerative changes in hair that start soon after the hair shafts grow out of the hair follicles, and become worse and worse towards the distal ends, or tips, of the hair.

Healthy, undamaged hair is soft, resilient and easy to detangle. The intact cuticle, the outer most layer of the hair shaft, gives the hair strength, shine, smoothness, softness, manageability and, most important, covers the cortex of the hair shaft which contains the structural elements and pigments of the hair filament. Sebum, the waxy, oily secretion from the sebaceous glands, is the natural, physiological coating that protects the cuticle of the hair shaft.

Traumas, such as shampooing, drying, styling, chemical dyeing and permanent waving, as well as exposure to sunlight, air pollution, wind, seawater, chlorinated swimming pool water and cigarette smoke, all take their toll on luxuriant hair. These traumas cause breaks, missing or denuded regions of the cuticle, and expose the cortex of the hair shaft to permanent, regional damage. Fractures as both transverse and lengthwise breaks (split ends or, medically, trichoptilosis) occur in the hair shafts. The altered structures of the weathered hair shaft increase the likeliness of further damage by the friction caused by normal combing and brushing. We love our hair, but we treat it brutally.

Proper selection of shampoos and conditioners can to some extent mitigate hair damage by decreasing brittleness, covering breaks in the cuticle, improving sheen and increasing strength. However, hair care products cannot stimulate the body's repair of weathered hair because the hair shaft is not a living structure and, therefore, actual repair of hair does not occur.

CONSUMER PRODUCTS

The hair care industry has developed a wide variety of products to make your daily chore of caring for your hair easier, more effective and rewarding. The hair care industry has also made a lot of money from your daily chore.

Shampoos

The English word shampoo is derived from the Hindi word *champo* which means head massage, usually with some form of oil. The enterprising Dean Mahomed introduced this therapeutic massage to London early in the 19th century and was soon appointed Shampooing Surgeon to George IV and then William IV. In the 1860's, shampooing took on a new meaning related to applying soap to the hair of the scalp. Actually, the use of soap for washing hair dates back millennia to ancient times. The problem with using soap is that it leaves hair dull and is often irritating.

Soaps, detergents and surfactants are all related chemicals. Some are naturally occurring chemicals, and are extracted from animal and plant material. Many more are synthesized in industrial laboratories. These chemicals are all relatively long molecules (visualize an alternating set of sticks and connecting balls, about 15-20 sticks long) that have a fat loving head (lipophilic) and a water loving tail (hydrophilic). Oils, grime, grease and dirt all tend to be like fats so they adhere to the fat loving parts of soaps, detergents and surfactants while the water loving parts of the molecules whisk them away in copious amounts of rinse water down the drain. Think of these chemicals as crocodiles with their head and forearms on the riverbank and their abdomen and tail in the water. The mouth of the crocodile grabs a passing baby gazelle and then, with a backward movement of the tail, the crocodile slips into the water.

The choice of whether to use a soap, detergent or surfactant is dependent upon the material to be washed and the desired outcome. Usually, soaps are for hands; detergents are for laundry; surfactants are for hair. Generally, soaps are the harshest.

The first modern shampoos with man-made chemicals were introduced in 1933 and are now the most widely used hair care products. Shampoos remove the natural sebum that has become dirty and greasy from perspiration, environmental dust, chemicals and dead cells. The trick is not to remove all of the sebum that makes hair shiny and manageable. Shampoos can enhance the natural beauty of hair and cover up, to some extent, the damage often inflicted on hair by the person with the blow-dryer, heated curlers, etc. The ultimate formulation of a shampoo is a mixture of many chemicals but shampoo commonly contains primary and secondary surfactants to produce foam and to act as detergents. Lots of foam is expected by people when they wash their hair. Consumers equate abundant, long-lasting foaming ability with cleansing ability, although, technically, excessive bubbles are not necessary for good cleaning. Baby shampoos are not irritating because they have very low amounts of surfactants. Babies do not produce much sebum.

If you look at the ingredients in the small print on the side of a bottle of shampoo, you will be overwhelmed by chemistry, and rightfully so. But, not to worry. Most of the chemicals are either natural, meaning they can be found in nature (after laborious extraction with you-don't-want-to-know-what), or chemicals synthesized in laboratories but based on natural molecules. In addition to the surfactants, which are the cleansers, there are chemical additives for special purposes and aesthetic effects, such as thickeners, opacifiers, emulsifiers, softeners, antioxidants, preservatives and fragrances. Most of these chemicals have been more or less tested for safety and can be found on the FDA's GRAS list, meaning they are "Generally Regarded As Safe." Other ingredients that find their way into shampoo via the imaginations of the marketing staff of hair care companies, such as oils from exotic plants, vitamins, keratins, collagens, amino acids and sun blocks, have limited, if any, usefulness.

Shampoos for oily hair have very effective cleansing surfactants or detergents and minimal conditioning agents. These shampoos are produced for use by adolescents or people with dirty hair. For dry hair, there are moisturizing shampoos. These shampoos contain mild detergents for cleansing, and various oils and sometimes silicone or petrolatum for conditioning. After shampooing, the hair feels wetter, smoother and more free-flowing. Moisturizing shampoos detangle hair, reduce static electricity and leave hair easier to comb. These products are aimed at use by people with weathered hair, dyed hair and African type hair.

Medicated shampoos for dandruff and parasite infestation contain additives such as tar derivatives, salicylic acid, selenium, chlorinated compounds, zinc salts and various anti-bacterial or anti-fungal agents. In some medicated shampoos, menthol is added to produce a tingling sensation that patients find reassuring.

Alternatively, there is the "No Poo" movement. Proponents point out that shampoos are expensive, irritating, not natural and that the removal of sebum by shampoos causes the sebaceous glands to produce even more sebum. In other words, your hair gets greasier and further hooks you on the products from the big, hair care products companies. You can buy no poo products or you can mix up a solution of baking soda and apple cider vinegar, and treat yourself to a no label, no poo shampoo.

Conditioners

Although not used as often as shampoos, conditioning agents are critical for sustaining the integrity of hair, particularly when hair is colored, weathered or aged. The complex mixtures of chemicals in conditioners are designed to improve the texture, appearance, shine and manageability of hair. Conditioners also increase the water content of each hair shaft.

Natural oils, such as tea tree oil and jojoba oil, have been used for centuries to condition human hair. Macassar oil was used by Victorian gentlemen, but it was so greasy that a small cloth was pinned to sofas and chairs to prevent staining of the upholstery. At the 1900 Exposition in Paris, Edward Pinaud introduced Brilliantine as a less greasy alternative and smoothed the way to the development of modern conditioners.

Conditioners work by sealing damaged regions of the hair shaft. They contain polymers or silicone derivatives that form a film, coating each hair. These films provide protection against thermal and mechanical damage, add volume, enhance body and improve resistance to breakage. Combing and brushing hair causes individual hair shafts to become negatively charged. Because they are relatively acidic, conditioning agents can eliminate static electricity making hair more manageable. Some conditioners claim to strengthen hair by providing hydrolyzed proteins to diffuse into the hair shaft. However, the added proteins readily diffuse in and out of the hair shaft, and the effect is slight, if any. Humectants, such as panthenol, increase the retention of water in the hair shaft giving each hair more volume. Some conditioners are engineered to produce specific results, such as:

- cream rinses have lower amounts of the chemicals that are in regular conditioners so as not to build up residues.
- pack conditioners, which are heavy and thick, glue the surface scales of the hair shaft together.
- hold conditioners use electrically charged chemicals to help the hair retain a styled shape.
- leave in conditioners are designed to maintain the hair soft and free flowing.

However, too much conditioning can make hair less manageable, and appear limp and oily.

The proteins in hair can be damaged by UV rays from the sun, causing dryness, reduced strength, rough surface texture, loss of luster, stiffness and brittleness of the hair shaft. There are also changes in color; brunette hair tends to develop reddish hues and blonde hair develops photo-yellowing. Many conditioners now contain photo-protection chemicals that have SPF ratings in an attempt to limit the action of UV rays and the subsequent sun damage to hair. However, it is likely that these chemicals are too dilute when applied to the hair and then rinsed out, meaning they do not have much effectiveness.

HAIR CARE, TRICHOLOGY, FUNGI AND NITS

FOOD IN HAIR AND HAIR IN FOOD

When you were a toddler, your parents watched in dismay as you picked up mushy food from the high chair tray and, with a big smile on your face, smeared the food into your hair. As an adult, you are still smearing food in your hair.

Most hair products on the pharmacy shelves or available at hair salons have some kind of food in them. For example, shampoos and conditioners often include a selection of coconut oil, peppermint oil, avocado oil, extracts of fennel, rosemary, hops, mint and mistletoe, and the juices of blueberries, cucumbers and carrots. The effectiveness of any of these ingredients is in the mind of the user. There are also many do-it-yourself home treatments for hair that are usually available in the kitchen. These include eggs, beer, olive oil, baking soda, mayonnaise, honey, molasses, maple syrup and apple cider vinegar, to name a few. Some of these foods are recommended to be left in the hair in for 6-8 hours, but users of these home potions swear by them.

Whereas food in your hair is something you may have been doing all of your life, hair in your food is disgusting. The presence of hair in food carries a heavy stigma of unpleasantness. Views differ as to the level of risk that hair in food poses to the inadvertent consumer, but there is at least a small risk that a hair may induce choking and vomiting, and also that it may be contaminated by toxic substances. Often, hair in food is interpreted as a sign of problems with hygiene at a restaurant.

In most developed countries, people working in the food industry are required to cover their hair. The introduction of complete capture hairnets has resulted in a decrease in incidents of contamination of food with hair. Also, in most restaurants, waiters are not permitted to have beards. When people are served food that contains hair in restaurants, they complain to the waiter. For the irate diner, hair in food has not been valid grounds on which to sue the restaurant in the US, unless the person is clearly injured by the hair in the food. Usually the matter is quickly handled by the staff with apologies, replacement servings and, perhaps, a free dessert.

HAIR AND DIET

Eating a healthy diet is an important part of maintaining healthy hair. The cells in the hair follicle are working constantly and very hard to produce protein rich hair; they need a complete supply of nutrients to support their activity. Poor nutrition takes a toll on many body functions and making hair is one of them. For example, protein deficiency can lead to brittle hair and a very low fat diet can diminish the natural luster of hair.

There are some foods that are particularly good to eat in order to have healthy hair. The omega-3 fatty acids in salmon will give hair a shiny look. Dark green vegetables like spinach, broccoli and Swiss chard supply vitamins A and C, which are needed to produce the sebum that coats healthy hair. Of course, carrots are a good source of vitamin A. Legumes like kidney beans and lentils supply iron and biotin that promote growth and strength of hair. Eggs also provide biotin as well as vitamin B12. Nuts, such as Brazil nuts, walnuts, cashews and almonds, contain zinc that will keep the scalp healthy and may reduce the number of hairs that are shed. Whole grains and oysters are also rich in zinc. And, a good source of protein is needed, such as fish, chicken, turkey or meat.

Be careful with those fad diets to lose weight, because dieting can lead to less than healthy hair as well as a noisy stomach. Many weight loss diets are deficient in zinc, omega-3 fatty acids and vitamin A, and may stunt hair growth, decrease hair shininess and even lead to hair loss. It would be a shame to lose a few pounds and end up with lackluster hair. Take supplements when dieting.

BUBBLE HAIR

When hair is heated, some of the water inside the hair is turned into steam. Hair dryers, irons and curlers operate between 250° F and 350° F. Water boils at 212° F. Chickens roast at 375° F. If a hot curling iron is put onto wet hair, it boils the water inside the hair. The steam from the boiling water expands and forms tiny bubbles, which become gas filled cavities inside the hair shaft.

Bubble hair is an acquired, self-inflicted deformity of the hair shaft that is associated with the use of hairdressing equipment that produce high heat. The degree of bubbling, therefore, increases with higher levels of moisture content and the higher heat generated by the hairdressing accessories. All hair will react this way to intense heat and the structure of the hair shaft is weakened resulting in marked fragility. There is no way to fix bubble hair, only to cut off the parts of the hair containing bubbles and to wait for the hair to regrow.

PRODUCT DEVELOPMENT AND TESTING
Researchers in the laboratories of hair care companies have developed ingenious methods to test products under development using objective methods to demonstrate efficacy. Measurements are made on single strands of hair or groups of hairs. Robotic and computerized systems are used to quantify cosmetic phenomena, such as build up, bounce, body, manageability and entanglement. Causes of damage can be modeled in the laboratory, as can the benefits of treatments, for example conditioning. When aiming at a global market, different hair types (European, Asian, African) are tested. Also, climate chambers are used to simulate dry, damp, hot and cold weather. Although laboratory test results can be difficult to extrapolate to a whole head of hair, the research data obtained are considered valuable by the industry for the further development of a product.

THE PROFESSIONALS
If things up there get really bad, you may want to see a trichologist. A trichologist is a health care professional who specializes in diseases that affect the hair and scalp. There are several professional organizations that focus on the scientific and medical aspects of trichology, such as: The Institute of Trichology, The International Association of Trichologists and The Trichology Society.

Trichology has its roots in England. Today, there are trichology centers in England, France and Italy. In the US, there are high concentrations of trichology centers in the "looking good" centers of the country, the Los Angeles and New York City areas. Trichologists in the US often claim to have been trained in Europe and many offer alternative medicine treatments. However, on their websites, the science can be faulty. There are no medical or clinical training requirements to call oneself a trichologist, so a trichologist may or may not be a physician.

Can you imagine introducing yourself at a cocktail party as a trichologist? Nobody would know what you are. And the puns! Trichologists are not magicians but they may be able to help you with your hair problems.

Nevertheless, if you are worried about the health of your hair, you will want to see a dermatologist or, at least, your family doctor or for your children, a pediatrician. Dermatologists, skin doctors, have the most training, which is not much, for dealing with health issues related to hair. Of course, specialists in plastic surgery and hair transplantation can help with the aesthetics.

DANDRUFF

Dandruff is in hair but is not actually a product of hair. Dandruff (medically pityriasis simplex capillitii) is the shedding of dead skin cells from the scalp. As it is normal for skin cells to die and slough off, a small amount of flaking is common and in most people, these flakes of skin are too small to be visible. Some people, however, either chronically or as a result of certain pathophysiological triggers, experience an unusually large amount of flaking, which can be accompanied by itchiness, redness, irritation and bacterial infection.

Dandruff is the result of three factors: sebum from the sebaceous glands, a metabolic by-product of skin micro-organisms (Malassezia yeasts) and individual susceptibility. Malassezia yeasts are a type of fungus. The scalp specific fungus, *Malassezia globosa*, is a normally found resident member of the microflora living in the scalp and, when triggered to grow in excess, is the responsible and treatable bug causing dandruff. There are about ten million of these fungi on your head right now.

Malassezia globosa metabolizes triglycerides, present in sebum, by the expression of lipase enzymes, resulting in a lipid by-product, oleic acid. Penetration of excessive oleic acid into the upper layer of the epidermis, the stratum corneum, causes an inflammatory response in susceptible persons that disturbs homeostasis, resulting in erratic cleavage and significantly increased sloughing of the top layer of dead cells. Once shed, these dead cells work their way through the hair and drop onto clothing. Dandruff flakes are oily clumps of dead skin cells, which appear as small white or grayish patches on the scalp, hair and black cashmere sweaters or blazers.

In any one individual, the trigger for dandruff to occur may not be clearly known and people differ in their susceptibility. There is some evidence that food (especially sugar and yeast), excessive perspiration and climate have significant roles in the pathogenesis of dandruff. Rarely, dandruff can be a manifestation of an allergic reaction to chemicals in hair gels/sprays/shampoos.

Those affected by dandruff find that it can cause social or self-esteem problems; treatment is important for both physiological and psychological reasons. Most cases of dandruff can be easily treated with medicated shampoos, containing selenium disulfide or ketoconazole to bring the Malassezia yeasts under control. Other treatments include ciclopiroxolamine, coal tar or zinc pyrithione medicated shampoos. Hydrogen peroxide is also used to manage symptoms of itching, although it can bleach the hair.

DERMATOPHYTES

Ringworm, although that is a misnomer, is indigenous to about 5% of the urban population in North America. Ringworm of the scalp, tinea capitis, is a fungal infection of the hair on the head that occurs mostly in young children. The causative organisms are members of either the

Trichophyton family or the *Microsporum* family of fungi. They are not worms. These fungi are anthropophilic, meaning they love people. What they particularly love is to eat hair.

The fungus invades the hair follicle and sets up home between the inner wall of the follicle and the hair shaft. Then, the fungus sends out long, slender, filamentous hyphae, which are branching tentacles of oozing cytoplasm that grow into the hair shaft. With its slimy hyphae in the hair shaft, the fungus makes mush of the inside and dines on the structural proteins of hair by making enzymes to digest the keratin. When the fungus eats up all of the keratin, the hair shaft collapses and breaks off at the level of the scalp. Usually, there is some mild inflammation associated with the de-haired site, which produces the red ring that is characteristically seen on the surface of the skin of the body, but may be hard to see on the scalp. The hyphae produce spores (arthroconidia) for further dissemination causing ringworm to get worse as the rings get bigger.

The fungus can be observed in the hair of a patient by using Wood's lamp. Essentially, Wood's lamp is an ultraviolet bulb that causes chemicals made by the fungus to fluoresce blue or green.

Treatment of tinea capitis is usually griseofulvin, an anti-fungal drug, taken orally for 4-8 weeks. Topical anti-fungal drugs are not very effective because they do not penetrate the hair follicles on the head to kill the growing part of the fungus. Newer anti-fungal drugs, such as terbinafine and the triazoles (itraconazole, fluconazole and voriconazole), can be effective and have the benefit of a shorter treatment period.

Another disease caused by fungi that invade the hair shaft is piedra. Piedra comes in two forms: black piedra is more common in hot, humid, tropical climates; white piedra is more common in semi-tropical or temperate areas. Piedra causes nodules to form around the hair shaft. Because anti-fungal drugs have limited success against piedra, the usual treatment is shaving the scalp.

LICE

People have been dealing with lice for so long that their creation is even mentioned in the Bible. As is told, Aaron struck the dust of the earth with his staff and the dust became lice for people and animals.

Although there are many species of lice, only one (*Pediculus humanus capitis*) infests the human head. In today's world, head lice are usually seen in young, school age children and affect approximately nine million people in the US each year across all socioeconomic strata. Transfer of lice directly from head to head occurs easily as children play together. Tools for use in hair, like combs and brushes, are not usually the source of the transfer from person to person. Although lice infestation does not pose much of a health threat, their very presence is a rich source of embarrassment to child and parent.

Lice are wingless insects, less than one tenth of an inch long. Lice cannot jump or fly, but they can crawl. They have three pairs of clawed legs that are designed with a sort of opposing thumb to grasp hair. In case you are interested, male lice can easily be distinguished from female lice. The males are smaller and have brown bands traversing their abdomen; the longer females have a larger abdomen.

Lice can live up to 4 weeks, during which time male and female lice copulate in hair. Copulation takes about an hour. Young males may copulate with older females but those old gals die soon after the act. A young fertilized female can lay 3-10 eggs/day. The eggs are enclosed in a casing, called a nit, which is firmly attached to the hair shaft. The eggs hatch nymphs in 10-14 days that then eat and grow for about 10 days, mate and the cycle continues.

Lice are very well adapted feeding machines. They pierce the skin of the scalp with their retractable mouth, a serrated tube-like protuberance (haustellum), and inject saliva containing vasodilators and anticoagulants. They like blood fresh; they do not like clotted blood. They eat every 4-6 hours and cannot survive away from the head for more than 24 hours. After having lice for a while, a person may develop a slight inflammatory reaction causing irritation and itching.

HAIR

Lice may be detected by visual inspection of the hair but are more likely to be seen when a fine toothed comb has been systematically used to comb the hair. The nits persist on the hair shaft for some time. Seeing nits means that an active lice infestation will recur.

Lice infestation is treated with creams that contain permethrin or malathion. These chemicals interfere with the ability of the lice to breathe. The creams have been very effective, but, alarmingly, lice are developing resistance to the active ingredients. A host of other drugs, both topically and orally, have been used against head lice with mixed success. Many experts recommend nit picking and avoiding the use of medications. Probably, the best way to deal with lice infestation and transfer to others is screening by parents or school personnel. Some schools have established a "no nit" policy.

Genetic data about lice has contributed to the field of archeogenetics. Scientists who studied lice have discovered some important information about human evolution. A significant change in the genetics of lice occurred around 170,000 years ago which led to two varieties of lice to evolve that are almost identical but do not mate. One variety of lice likes the environment of the human head and the other variety of lice likes the environment of the human body. Why did two varieties of nearly identical lice appear 170,000 years ago? Evolving humans started to wear clothes. Human head lice do not like to be under clothing.

FLEAS

You can get fleas in the hair on your head, but you usually don't. Fleas are like lice, but bigger. Fleas are ugly insect critters that instead of flying have developed an incredible ability to jump. They jump from furry mammals like pets or farm animals to other furry species. Occasionally, they jump onto humans. Like lice, fleas want your blood. Usually, their jump lands them on your legs or arms, and then you get flea bites. Fortunately, they cannot jump as high as your head, but they can get into your bed.

If you get fleas in the hair on your head, a few rounds of shampooing will get rid of them. Because fleas are not really a problem for the hair on your head, we will not delve further into their repulsive lifestyle.

We have discussed how we take care of our hair with shampoos, conditioners and perhaps a visit to the trichologist or dermatologist when creatures invade. Now we return to our initial definition that differentiates hair from fur. Hair needs to be cut and the vast majority of people have their hair cut often. Let's look at what it takes to be a cutter of hair.

10

CUTTING HAIR AND BLOODLETTING

"Doth not even nature itself teach you, that if a man have long hair, it is a shame unto him?"

-Paul, Corinthians

Most people in the world cut their own hair or have it done by a spouse, relative or friend. Most men in the world shave their own facial hair or trim their own beard. In other words, hair is there and when it gets too long, it gets cut by the hairee or a member of the tribe. In many ethnic groups, family and close friends style and adorn hair for ceremonial occasions according to cultural traditions. Haircutting, styling and decorating is commonly, throughout the world, something done at home.

But, that's the *rest* of the world. In western civilizations the cutting, styling and coloring of human head hair has become a very popular, profitable and competitive profession. In typical, small, American towns, it seems like the collection of barbershops, beauty parlors and hair salons often outnumbers the collection of pharmacies, dry cleaners and pizza joints altogether.

MY BARBER

Having suffered male pattern baldness all my adult life, my haircuts are not a big deal. However, I did not know how one becomes a barber. So one day, I stretched a five minutes haircut into a thirty minutes interview with my barber.

Maryanne is stout and of medium height, about 50 years old and has long black curly hair. She works in an old fashioned barbershop in a small town in New Hampshire. The shop has two barber chairs and an L-shaped area with cushioned metal chairs for waiting men and boys. There are local newspapers and old magazines available to pass the time while waiting. When one of the barbers finishes cutting the hair of a client, somehow it is known: "Who's next?"

Maryanne is the third generation of barbers to service this town. Her grandfather and father were barbers, and now she and her sister have taken over the family business. Her sister has always been a barber. Maryanne retired after 20 years of being a police dispatcher and became a barber in 2002.

To become a licensed barber, Maryanne went to barber school where her training included haircutting and hairstyling. At the time, there were no barber schools in New Hampshire so she was trained in Massachusetts. To be licensed in New Hampshire, Maryanne had to have 1700 hours of training. That is almost 10 months of full time, five days a week schooling. For comparisons, police academies turn out new cops in 4-5 months, dental hygienists are trained for 48 months and journeymen plumbers carry the heavy equipment for master plumbers for 4 years.

Professionals who cut hair have usually attended an accredited school where the training costs $6,500-$10,000. The largest accrediting agency for barber and cosmetology schools is NACCAS, which stands for the National Accrediting Commission for Cosmetology Arts & Sciences. NACCAS is not a government agency, but the US Department of Education recognizes their authority in accrediting schools that teach haircutting and styling. NACCAS currently reviews approximately 1,000 institutions that serve over 100,000 students. These schools offer more than 20 courses and programs of study, which fall under the NACCAS scope of accreditation. There are several other such agencies with similar guidelines.

Current licensing requirements in New Hampshire require 800 hours (about 5 months) of training and, to be a barber now, there is no need any more to learn styling. Requirements are set by each state and can vary, differ and change significantly from time to time. Both written and practical exams are usually required, as regulated by a State Board. Many states still call for about 1,500 hours of training to be a barber. In most states, a person must be at least 16 years old and have graduated from an accredited barber or cosmetology school to be a professional cutter of hair.

While at barber school, the first time Maryanne held a scissor and cut hair was on a real person. She also learned how to shave a man with a straight edge razor. "No balloons" she told me. "We got a lot of bums who would come in once a week for a shave."

BARBERS, HAIRDRESSERS, HAIRSTYLISTS, BEAUTICIANS AND COSMETOLOGISTS

A barber (from the Latin *barba*, which means beard) is someone whose occupation is to cut any type of hair. In olden days, because shaving required mastering the techniques of sharpening and using a straight razor, barbers were most often the ones to shave or to trim beards. In more recent times, with the development of safety razors and the decreasing prevalence of beards, most barbers primarily cut men's hair. Barbers do not commonly offer significant styling or cutting edge haircuts, but some barbers still deal with facial hair when requested and can style artificial hair replacement products, such as toupees, usually in the back room. The place where a barber works is generally called a barbershop and often has an old fashioned, Main Street feeling. The average annual salary for a barber in the US is $25,000 with a large variation based on location, experience and clientele. Tips are very important.

HAIR

A hairdresser is a general term, first used in 1771, referring to someone whose occupation is to style hair, usually women's hair, in order to maintain or change a person's image. The hairdresser uses a combination of hair coloring, haircutting and hair texture techniques. In terms of haircutting and styling, the differences between hairdressers, hairstylists, beauticians and cosmetologists are minimal. Training for the different titles requires from 1,000 hours to over 1,500 hours. Although the diploma may be different, all must graduate from an accredited school and be licensed by the state in which they practice. Like a barber, the average annual salary for a hairdresser, etc. in the US is $25,000 but varies with location, experience and clientele. Again, don't forget to tip.

Today, hair professionals work side by side in establishments known as hair salons, which tend to be much more chic than barbershops. Hair salons seek to accommodate the latest hairstyles by combining traditional haircutting with skill, daring, and modern practices and equipment. Hair salons for men have afforded the barber the opportunity to be progressively contemporary. In hair salons for women and unisex hair salons, the cutting, coloring and styling of hair in the latest fashions and trends may be done by people with a variety of different titles. In 2007, there were about 1.7 hair professionals working in barbershops and hair salons in the US.

On another note, the barbershop was the founding location for an endearing, harmonious kind of popular music: The Barbershop Quartet. The lyrical, nostalgic songs are sung *a cappella* by a group of four men who reach an acoustical overtone and produce ringing chords with their four part harmony. This style of singing comes from southern roots circa mid-1800's when African-American quartets "cracked a chord" at, for example, Joe Sarpy's Cut Rate Shaving Parlor in St. Louis. There was a revival of interest in the 1940's and today, this sentimental type of singing continues with all male barbershop quartets, with all female barbershop quartets and at an annual, international competition in Kansas City.

EARLY HISTORY OF HAIRCUTTING

Cutting off hair, sometimes all of the hair, is an ancient practice. Some primitive people believed that bad spirits entered the body through hairs and therefore cut them off, as best they could. Haircutting tools have been found among relics of the Bronze Age, circa 3,500 BCE. In Egypt, old monuments and papyruses show pharaohs having their head shaved and Egyptian priests were de-haired every three days. Probably, this was a practical way to keep the head free of parasites. In the Bible, Ezekiel commands: "Now, son of man, take a sharp sword and use it to shave your head and your beard." In the 5th century BCE, barbers worked near the ancient Greek houses of government, trimming and styling the patron's hair and cutting fingernails, as philosophical debate flowed freely.

Barbering was introduced to Rome through the Greek colonies in Sicily about 300 BCE, and barbershops quickly became very popular centers for daily news and gossip. A morning visit to the barbershop became part of the daily routine, as important as a visit to the public baths. A few Roman barbers became wealthy and influential, running shops that were favorite public places of high society. However, most were simple tradesmen, owning small storefronts or setting up their stool in a street near the Forum and offering haircuts and shaves.

BARBER-SURGEONS

During the Middle Ages in Europe, the barbers had the sharpest knives and, therefore, were important adjuncts to the medical community as surgeons. In addition to haircutting and shaving, medieval barbers performed surgery, amputation, bloodletting and leeching, fire cupping, enemas, and the extraction of teeth. Other reasons to go to the barber included neck manipulation, cleaning of ears and scalp, draining of boils or fistulas, and lancing of cysts with wicks. Barber-surgeons were very much a part of the practice of medicine, and made house calls to the wealthy and had small dingy shops in the villages for the locals.

Bloodletting, a common treatment for diseases for millennia, was prescribed by physicians but carried out by barbers with their honed blades. Patients would tightly grasp a rod or staff so that their veins would show and barbers would cut the veins allowing blood to flow out until the patient fainted. Often, a patient was treated repeatedly over several days with bloodletting. The use of leeches, also administered by barbers, was considered an alternative to the cutting of the blade.

The rationale for bloodletting was the widespread medical belief, dating back to Hippocrates, that diseases were caused by an over (or under) production of certain body humors. Blood was one of the four body humors and certain diseases occurred when, for example, the body contained too much blood. Bloodletting was used to treat acne, asthma, cancer, cholera, coma, convulsions, diabetes, epilepsy, gangrene, gout, herpes, indigestion, insanity, jaundice, leprosy, ophthalmia, plague, pneumonia, scurvy, smallpox, stroke, tetanus, tuberculosis and for some one hundred other diseases. Bloodletting was even used to treat most forms of hemorrhaging, such as nosebleed, excessive menstruation or hemorrhoidal bleeding. Before surgery or at the onset of childbirth, blood was drained to prevent inflammation. Before amputation, it was customary for barbers to remove a quantity of blood equal to the amount believed to circulate in the limb that was to be sawed off.

To protect the trade from just anyone setting up shop, guilds were formed in France and England. The first known organization of barber-surgeons was founded in 1094 in France. In 1308 in Great Britain, the Worshipful Company of Barbers was established and still exists. It wasn't until the 1700's that both France and England began to prohibit barbers from performing surgery. Not so in the New World where barbers wielded their razors, and were the surgeons and dentists until more recent times. The first President of the United States, George Washington, died in 1799 after 5 pints of blood were removed from his body in 24 hours by a barber-surgeon as treatment for a throat infection. By 1900, bloodletting was considered quackery in the US.

THE BARBER'S POLE

The iconic sign marking a barbershop is a twirling cylinder with spiraling red, white and sometimes blue stripes outside the front door. The barber's pole of today is laden with symbolism related to barbers having done bloodletting hundreds of years ago. The red and white stripes symbolize the bandages used by the barber-surgeon during the procedures, red for the blood stained and white for the clean bandages. The cloth bandages were literally hung on a pole outside of the shop to dry after washing. As the bandages blew in the wind, they would twist together to form the spiral pattern. Those twirling bandages were how patients in medieval Europe knew where to find the barber-surgeon.

At some point, the cloths were replaced by a wooden pole that was painted with red and white spiraling stripes. A brass cap at the top represents the vessel in which leeches were kept, and a brass basin on the bottom represents the bowl that received the blood. The pole itself can be thought of as the staff that the patient gripped during the procedure to encourage blood flow. Spinning barber's poles are supposed to be oriented so that the blood (the red stripes) will appear as if it is flowing down into the bottom basin.

In the US, the barber's pole is often red, white and blue. Red represents arterial blood, blue is symbolic of venous blood, and white depicts the bandages. Or maybe it is just patriotic. Prior to 1950, there were four manufacturers of barber's poles in the US. In 1950, William Marvy of St. Paul, Minnesota, started manufacturing barber's poles and by 1996, over 74,000 barber's poles had been produced by the William Marvy Company, which is now the sole manufacturer of barber's poles in North America. In recent years, the sale of spinning barber's poles has dropped considerably; fewer barbershops are opening or are being replaced by hair salons.

NOT TO BE CONFUSED WITH...

In some parts of Asia, a red, white and blue barber's pole is used as a symbol for a brothel. Whereas prostitution is illegal in many parts of

Asia, laws against it are often not enforced to the degree that all public solicitations are eliminated. The barber's pole is used as an understood way of identifying houses of prostitution.

In South Korea, for example, brothels disguised as barbershops sometimes use two poles next to each other, often spinning in opposite directions. Actual barbershops try to distinguish themselves as places that cut hair and will usually use one pole that shows a picture of a woman with flowing hair and the words "hair salon" written on the pole. That way, the person can be sure that the service that they are paying for is to get a haircut.

FAMOUS BARBERS

In real life, barbers become famous when they cut, color and style the hair of movie stars, presidents and sports deities. The cost of getting done by one of these hair specialists is very high. In fiction, barbers become famous because they are such notorious characters.

Sweeney Todd, who has been a psycho-barber in newspaper serials, theater and movies, dispatches his victims by pulling a lever while they are in his barber's chair. The victims fall backward down a revolving trapdoor into the basement of his shop, causing them to break their necks or crack their skulls. Just in case they are alive, he goes to the basement and polishes them off by slitting their throats with his straight edge razor. In some adaptations, the murdering process is reversed, with Todd slitting the throats of his customers before they fall into the basement via the trapdoor in the floor.

After Sweeney Todd has robbed his dead victims of their valuables, Mrs. Lovett, his partner in crime and perhaps his lover, assists him in disposing of the bodies by baking their flesh into meat pies, and selling them to the unsuspecting customers at her pie shop. Todd's barbershop was supposedly located at 186 Fleet Street, London, next to St. Dunstan's Church, and was connected to Mrs. Lovett's pie shop in nearby Bell Yard by an underground passage. The tale surrounding the character was a staple melodrama of Victorian England, but claims that Sweeney Todd was a real person are strongly disputed by scholars.

Another famous barber is Figaro, in Rossini's comic opera *The Barber of Seville*. Opening night in 1816 in Rome was a disaster, made so by the shenanigans of supporters of a competing operatic composer. Nevertheless, the audience soon came to love the silliness of *The Barber of Seville*, which has remained a popular opera through the ages.

Figaro was always ready to meddle, particularly for money. He befriends the Count who has fallen in love with Rosina, whom Dr. Bartolo has taken in as his ward and intends to marry. Figaro disguises the Count as a crazed soldier to get him into Dr. Bartolo's house to see Rosina. Later, while giving Dr. Bartolo a shave, Figaro steals the house keys so that the Count can come for Rosina and take her away with him. Of course, everything gets mixed up and melodramatically funny, but the opera ends well with the Count and Rosina married, and Figaro handsomely rewarded. Figaro also appears in Mozart's *The Marriage of Figaro*, not as a barber but as the servant head of the Count's household.

FAMOUS HAIRCUTS

Samson's mother received a visit from an angel, who told her she would give birth to an extraordinary son and not to cut his hair. When he grew up, Samson had incredible strength. He purportedly killed a lion with his bare hands and allegedly slew 1,000 of the enemy Philistines with a jawbone of a donkey. Samson did have a weakness: Philistine women.

Samson had known many Philistine women, in the biblical sense, but he fell in love with Delilah. The leaders of the Philistines went to Delilah and paid her to seduce Samson and find out what made him so strong. When Delilah asked Samson where his strength came from, Samson told her that he would lose his strength if she tied him to the bed with ropes. After several rounds of bondage, Delilah used all her womanly powers to finally get Samson to tell her that his strength was because of his hair, which had never been cut. That must have been quite a night of passionate lovemaking because while he slept, Delilah ordered a servant to cut off Samson's hair. He woke up in the morning to Delilah's betrayal. Having lost his strength, Samson was captured by the Philistines, who gouged out his eyes and made him grind grain in prison.

During a celebration attended by thousands of Philistines, Samson was chained in the center of their temple, between the two main pillars. Samson asked God to strengthen him one more time, "So that I may pay back the Philistines for the loss of my eyes." Then Samson pushed against the pillars with all his might. "Let me die with the Philistines," he prayed. As the temple crashed down, Samson killed more Philistines at the time of his death than during his entire lifetime.

Bill Clinton got a haircut that made coast-to-coast news. In May 1993, Air Force One sat for two hours on a runway at Los Angeles airport as Christophe of Beverly Hills gave the President a razor cut. Because of the presence of the President, two LAX runways were closed and flights were reported to be delayed. Newspapers across the country called it "the most expensive haircut in history." As was his style, Clinton just shrugged it off. An FAA report that came out later indicated, in fact, that no planes had been delayed while the President was getting coiffed.

LOCK OF HAIR
A lock of hair has always meant a lot. In ancient Greek mythology, a forelock of hair was cut off a minor deity before being sacrificed by a major deity. In this way, the soul was released. In the early days of Islam, Muslims would allow a single lock of hair to grow on their shaven head so that Muhammed would have something to grasp in order to pull the deceased into Paradise. The lock of certain North American Indians, left on the otherwise bald head, was for a conquering enemy to seize when he tore off the scalp.

In more civilized times, a lock of hair has become a sign of love or a gift of endearment for the loved one to carry with him or her during travel, times of war or as a remembrance. In the 16th and 17th centuries, fashionable gentlemen wore a long, braided lock over their left shoulder to show devotion to a loved one. Sir Walter Scott wrote the poem, *To a Lock of Hair*, in remembrance of his lost love, Agnes, "If she had lived and lived to love me." During the Civil War, soldiers would leave behind a lock of their hair with their families and often carry a lock of hair of a loved one in a "locket" into battle. If the soldier perished in the war, the family might have had his hair woven into a piece of mourning jewelry, such as a brooch or a pin.

NOTES FROM THE FIELD
Sir James George Frazer was a famous Scottish anthropologist in the latter part of the 19th century/early 20th century. He was interested in myths, rituals, religions and folklore throughout the world and is considered the father of modern anthropology. However, he did not travel widely but gathered his information by letter correspondences with missionaries. In 1922, he published his collection of anthropologic tidbits regarding haircutting in primitive cultures in the far reaches of the globe. Whether these practices exist today is difficult to say unless someone has made a recent trip to any of the remaining tribes. Nevertheless, as reported at the time, here are some examples:

- In Fiji, the chief of the Namosi islanders always ate a man as a precaution before he had his haircut. A certain clan had to provide the victim, and the elders sat in solemn council to choose the offeree. There was a sacrificial feast to avert evil from the chief in case he got a bad haircut.
- Among the Maoris of New Zealand many spells were uttered at a haircutting. One, for example, was spoken to consecrate the obsidian knife with which the hair was cut; another was pronounced to avoid the thunder and lightning which haircutting was thought to cause.

HAIR

- The shamans of the Marquesan islanders in Polynesia took some of the hair of the man they wished to injure, wrapped it up in a leaf and placed the packet in a bag woven of fibers knotted in an intricate pattern. The packet was then buried and the victim was presumed to waste away from a languishing sickness that lasted twenty days. His life might be saved by discovering and digging up the buried hair. When they made a vow to avenge the death of a near relative, Marquesan tribesmen had their head entirely shaved, except one lock on the crown which was worn loose or put up in a knot. The lock was not cut off until the man had fulfilled his promise.
- Among the Toradjas in Indonesia, when a child's hair was cut to rid it of vermin, some locks were allowed to remain on the crown of the head as a refuge for one of the child's souls. Otherwise, the souls would have no place in which to settle and the child would sicken. Similarly, the Bataks in Indonesia were afraid of frightening away the soul of a child when they cut its hair. They always left a patch unshorn to which the soul could retreat. This lock remained uncut at least until manhood.
- When an Australian aborigine wished to get rid of his wife, he cut off a lock of her hair while she slept, tied it to his spear and took it to a neighboring tribe, where he gave it to a friend. His friend stuck the spear up before the nightly campfire and when the hair was roasted, it was a sign that the unwanted wife was near death.
- The Huzuls of the Carpathian Mountains in the Ukraine believed that if mice got a person's shorn hair and made a nest of it, the person would suffer from headaches or perhaps become an idiot.
- When the Nandi of Kenya captured a prisoner, they shaved his head and kept the shorn hair as a security that he would not attempt to escape. When the captive was ransomed, they returned his shorn hair with him to his own people.

- Turks and Armenians were said to take extreme care to preserve the hairs that were cut off or torn out with a comb so that the owner may have them at the resurrection of the body.
- The almost universal dread of hair being used in witchcraft caused various West African and South African tribes, the Tyroleans, the Patagonians and some Italian women to burn their cut hair. In the Solomon Islands, Tahiti and Melanesia, men buried their cut hair so that it did not fall into the hands of an enemy, who would make magic with it and bring sickness or calamity on them.

Returning to present day, the haircutting act is often performed at a hair salon, and there is no real thought about the cut hair that falls to the floor and is swept away. The focus is on the hair on the head in terms of length, color, contour and style. Much fussing is offered to the client who knows very little about the chemical and thermal engineering that is applied to her or his head. In the next chapter, let's observe what happens to the hair of Stephanie, a professional, stylish, upwardly mobile, thirty-something woman who is spending a morning at the hair salon.

11

A MORNING AT THE HAIR SALON

"I'm not offended by all the dumb-blonde jokes because I know that I'm not dumb. I also know I'm not blonde."

-Dolly Parton

Stephanie has great hair. She is a brunette with naturally curly, thick tresses. As a girl, Stephanie had spit curls and as a teenager, she could not make her hair do what she wanted it to do. There was much frustration. Now, she can jump out of the shower after her morning workout and drive to her office with her hair still wet. When she walks into the Board Room for the 9:00 AM meeting in her black Tahari suit, sits down, tosses her hair back and puts on her red executive glasses, she looks FABULOUS!

Stephanie's hairstylist, Roberto, has been a long-term trusted ally. He is part color artist, part sculptor, part chemical engineer, part thermal engineer, part mechanical engineer, part surgeon and part confidante. Through a series of trials and errors, mostly errors, Stephanie finally found Roberto. His sense of aesthetics matches Stephanie's self-image, so far. He is confident that he knows what she wants. He better be. After all, Stephanie has placed the topmost asset of her physical image in his hands. Not to mention a chunk of her last paycheck.

It is Saturday morning and in the evening Stephanie will be going to a semiformal event, related to the work of a new man that recently entered her life. She is at the hair salon to get done. Her hair will be colored, washed and styled. There is much work to do. Although Roberto regularly reminds his clients: "I am a beautician, not a magician," he can alter the unwanted effects of genes, of the environment and of all the daily-use hair care products, and produce a truly awesome head of hair. Tonight, Stephanie wants to look SPECTACULAR!

COLORING HAIR

Stephanie has been coloring her hair for the last few years for two reasons. She thinks that her dark brown hair should be lightened up a bit to complement her complexion and, perhaps more importantly, there has been the appearance of gray hairs. Less than half of women in their 20's color their hair, but more than 65% of women in their 30's add color on a regular basis. By mid 50's, most women are using the bottle. Very few of these women, including Stephanie, have any idea of what is going on when they submit their hair to chemical engineering.

Permanent coloring

Coloring hair is all about chemical reactions. Mixtures of toxic substances are applied to hair and through a series of chemical reactions, dyes, which are chemicals with colors, are formed and made to adhere to the hair. Chemical reactions are all around us but we do not usually associate them with hair.

We use chemical reactions in preparation of our food. By cooking meat and vegetables, heat causes chemical reactions that break down proteins, and make our food more chewable and digestible. Marinating with sauces causes chemical reactions that bring out sweet or savory flavors in our foods. Mixing flour, water and yeast causes chemical reactions that bring us our daily bread. There are lots of chemical reactions inside of our bodies that are the bases for metabolism, growth and health, but we do not think very much about them. When we speak about chemical reactions inside of our body, we call them biochemical reactions. Biochemical reactions make the pigments that give hair its natural color.

People have been using chemical reactions to change the color of their hair for thousands of years. Such was the work of groomers, skilled slaves, servants and ladies-in-waiting. In 1907, Eugène Schueller, a French chemist, invented a commercial hair coloring process based on using the laboratory synthesized, not natural, chemical *p*-phenylenediamine. He claimed safety for his man-made product. Success provided the foundation of his company, the French Harmless Hair Dye Co. Later, this company became L'Oréal.

The chemical reactions in permanent hair coloring products have functions that are designed to work in two steps:

1. *Chemically altering the hair shaft to maximize color change.*

The layers upon layers of tiny scales of the cuticle form a barrier to the inner parts of the hair shaft, the cortex, which is where the new chemical dyes need to be set. The cortex also contains the natural color hair pigments and any additional dyes from previous hair dying. To achieve permanent hair dying, the cortex must be made accessible to the dye chemicals. The hair coloring products contain hydrogen peroxide, ammonia and monoethanolamine to damage the cuticle and make hair fibers swell with water for better penetration of the added, color producing chemicals.

Virgin hair has underlying natural pigments to change but no artificial pigments; whereas, previously colored hair contains natural pigments as well as artificial pigments to deal with. An important aspect of coloring hair is to appreciate that color is being added to color and what to do about that. Coloring hair is not like a painter applying color to a blank canvas. When you walk into the hair salon, you bring in your original natural hair color or your dyed hair color. In particular, the coloring process does not always remove all previously placed artificial pigments, so new dye added to already existing artificially colored hair can give the desired color, something close to the desired color, or an entirely different, bizarre color.

The most drastic means to prepare hair for a new hair color is to remove all color with a bleaching agent. Hydrogen peroxide is commonly used. This chemical reaction literally changes the structure of the natural biochemical pigments, eumelanin and pheomelanin, so that they no longer are chemicals that have color. Such treatment produces hair that is featureless white or yellowish white and that can then be dyed. In addition, or alternatively, if previous dye is already present, a hair color removal product is used, which contains chemicals designed to remove at least some of the artificial pigments.

2. *Using chemical reactions to produce color.*

Stephanie is sitting in the chair when Roberto brings over two charts that are laminated because of their frequent use around water and chemicals. Before starting the hair color change process, there is an analysis. One chart is an international system of identifying the darkness of hair: 1 being the darkest (black) and 10 being the lightest (blonde). Most stylists are reluctant to change that number by more than 2-3 units. The other chart is like a color wheel that helps predict what color Stephanie will end up with when Roberto adds the new dyes to her current hair color.

For example, if hair is level 7 and you are trying to go lighter, you must use a purple-based color to neutralize the underlying yellow pigments. Darker than level 7, the orange underlying pigments come into play and a blue based color is needed. If hair is dark brown (level 3) like Stephanie's and she wants to lighten up to a more golden brown (level 6), the very strong red-orange underlying pigment needs a neutral or slightly cool, gold color to bring out some tonal brass.

This is a bit of a gamble. Stephanie is nervous. She does not want her hair stripped down to blonde. Too damaging. But by adding synthetic dyes to her natural color, will she get the color that she wants? Timing is critical; the hair coloring dyes cannot remain on her hair too long. Stephanie has heard stories of women ending up with green hair. Will adding too much gold color make her hair too light when Roberto is finished? Maybe she should be shooting for level 4 ½ or 5. Stephanie puts her trust in Roberto. He knows what he is doing. But, how do those chemicals that are about to soak Stephanie's hair actually work?

The dyes used to color hair are cousins of benzene, a product of the petroleum industry. Benzene is an unusual chemical with unusual properties because of its hexagonal structure. At different points around the hexagon, there can be long chemical side chains sticking out. Many of the chemical modifications to the benzene ring produce a spectrum of colorful molecules. Benzene-like colored chemicals are the bases for dyes used to make all kinds of colorful fabrics, paints, plastics and food colorings. In addition to being colorful, chemicals that are based on benzene often have a distinct smell and are referred to as aromatic compounds. They may look nice and even smell sweet, but benzene and its cousins can be quite toxic.

Permanent hair coloring products usually come as two components that are packaged separately and mixed together immediately before application. One package contains hydrogen peroxide in water or in a lotion base. The other package contains precursor dyes and preformed dyes in an ammonia solution. The precursor dyes are the benzene-like chemicals that develop color upon oxidation with the hydrogen peroxide. The preformed dyes, called couplers, react with the oxidized dyes to provide a wider variety of colors. There are blue couplers, red couplers, yellow couplers and green couplers. You have probably seen elderly women with blue or green hair; too much coupler applied to the already hard to dye gray hair. The couplers are related to other benzene-like chemicals and are used along with the oxidation dye precursors to add vibrancy and intensity. Darker shades are obtained by using higher concentrations of the precursors or preformed dyes.

The hair coloring process has traditionally been applied to the hair as one overall color. The modern trend is to use several colors to produce streaks or gradations, either on top of the natural color or on top of a single base color. Highlighting is done before the permanent base color by treating sections of hair with lighteners, usually to create blondish streaks. For lowlighting, sections of hair are treated with darker colors. There are different methods that can be used to produce these variations. Pieces of foil or plastic film are used to separate off the locks of hair to be colored, especially when applying more than one color. The coloring solution is placed within the foil and allowed to react with only the hairs sequestered in the foils. As another method, a plastic cap is placed tightly on the head, strands of hair are pulled through holes in the cap with a hook and color is applied manually. Or, hair color is painted directly onto sections of the hair by the hairstylist *artiste* with no foils used to keep the color contained.

For the base color, Roberto mixes the ammonia solution containing the dyes and the hydrogen peroxide solution together and applies the dark mixture immediately to Stephanie's hair. The ammonia in the concoction causes the cuticle scales to separate and her hair swells. Then, the dye precursors and couplers penetrate the cuticle, and stick to the cortex of the hair shaft. During the time that the hair coloring solutions are soaking Stephanie's head, she sits under a steam dryer. The steam moisturizes the hair which helps get the dye into the hair shaft where the full chemical reactions between hydrogen peroxide and the different benzene-like chemicals are going on. The oxidized products of the original precursor dyes are chemically attached to the couplers. All of these chemical reactions produce dyes whose chemical structures are very large, much larger than the initial precursor benzene-like chemicals. Because of the increased size, these large colored pigments get trapped inside the cortex of the hair shaft. Full color develops slowly over about 30 minutes and, hopefully, what Stephanie wanted is what she gets. When the time allotted for the coloring process is finished, these large colored chemicals are sealed into the hair shaft by the closing of the overlying cuticle scales.

All of these chemical reactions take place on the visible part of the hair shaft that has emerged from the hair follicle. The chemicals being used do not penetrate into the part of the hair shaft that is still within the hair follicle. The hair is permanently colored above the skin of the scalp, but the roots, as they emerge newly formed from the hair follicle, will have the original natural color.

And, how did Stephanie's hair color come out? Roberto took her dark brown hair, lightened it up slightly and gave it a vibrant overtone. Not a magician; just the daily work of a great beautician. And, a good, healthy head of hair to work on.

Semi-permanent coloring
Semi-permanent dyes penetrate into the hair shaft, but they are not as chemically reactive as permanent dyes. Semi-permanent dyes contain no, or very low levels of hydrogen peroxide and ammonia, and are therefore safer for damaged or fragile hair, but they cannot lighten hair. Although semi-permanent dyes do not readily rinse off with water, they do fade and wash out of hair after about 5-10 shampoos.

When using semi-permanent coloring, the final color of each strand of hair depends on its original color and porosity. There are subtle variations in shade across the whole head that gives a more natural result compared to the same all over color of a permanent dye. Gray or white hairs do not dye to the same shade as the rest of the hair. If there are only a few gray/white hairs, they usually blend in. But, as the gray spreads, there comes a point where it cannot be hidden as well. Time for permanent coloring, maybe even with some highlights.

Temporary coloring
Temporary hair colors are applied in the form of rinses, gels, mousses, foams or sprays. These hair color products contain chemical dyes that

coat the surface of the hair, the cuticle, but do not penetrate into the cortex of the hair. Because temporary hair color products are like a thin layer of paint on the scaly surface of the hair shaft and are not set into the internal structure of the hair shaft, these coloring agents usually wash out within 2-3 shampoos.

Alternative hair colorants

There are alternatives to industrial dyes and processes to color hair. Ancient civilizations, like the Egyptians, Greeks and Romans dyed their hair using natural pigments from plants, such as henna, indigo, cassia, senna, turmeric and emblic. Other pigments came from black walnut hulls, red ochre and leeks. Today, store bought products made with these traditional plant based colorants generally contain fewer potentially toxic chemicals. Available for purchase are temporary, semi-permanent and permanent treatments, but in practice the results obtained with these products often do not last as long as industrial strength dyes. Allergic reactions are possible even from natural plant dyes.

The orange dye, henna, is from the plant *Lawsonia inermis* whose active pigment, lawsone, binds to keratin. Depending on a person's hair type, henna can be semi-permanent or permanent. Most people achieve a permanent color from henna with repeated use as the orange color builds up into reddish brown or auburn.

Indigo is a natural dye from plants of the *Indigofera* family that can be added to henna or layered on top to create brown to black colors in the hair. On the color wheel, henna is orange, and indigo is blue, so the two colors together create brown tones. Indigo may fade after one application, but becomes more permanent with repeated use.

Using a plant-based color such as henna can cause problems later when trying to use synthetic based permanent hair color products. Many products sold as henna have been mixed with ingredients from other plants, dyes and additives that can lead to unpredictable results when the hair is subsequently colored with a synthetic dye. Although it may not be visible on darker hair, the staining from henna will remain for several months. This may only be realized when additional dying is attempted and an unpleasant, permanent orange color results. Be careful.

Alternative lifestyle hair colorants
Some rebellious young people like to color their hair shocking pink, day glow orange or bright purple. Don't worry, it will wash out, eventually. The chemicals used in alternative lifestyle color dyes typically contain tints but usually no color stripping chemical reactivity. However, these tints will only create the bright color of the packet if they are applied to light colored hair. Young people with medium brown to black hair will often use a bleaching kit prior to tint application. Even some people with fair hair trying to achieve bright pinks, blues and greens may feel they benefit from prior bleaching. Although alternative lifestyle colors are semi-permanent, some colors such as blue and purple may take several months to fully wash out from bleached or pre-lightened hair.

Adverse effects of hair coloring
In some individuals, the use of hair coloring may result in an allergic reaction and/or skin irritation. Symptoms of these reactions can include redness, sores, itching, burning sensation and discomfort. These responses may not be apparent soon after the application and processing of the dye, but can arise hours or even a day later. If any of these symptoms occur, the artificial dyes producing the color should be removed as soon as possible.

Hair that has been damaged by excessive exposure to chemicals is considered over-processed, resulting in dry, rough and fragile hair. In extreme cases, the hair can be so damaged that it breaks off entirely. Breakage is particularly a problem for African type hair, especially if used in combination with relaxers. Treatments are available but the only real solution is to stop the use of chemicals until new hair grows and the damaged hair is cut off entirely. Colored hair should be regularly washed and conditioned with gentle products specifically designed for color-treated hair. This will help keep the hair intact and minimize color fading.

Skin and fingernails are made of a similar type of keratin protein as hair. That means that drips, slips and extra hair tint around the hairline can result in patches of discolored skin. This is more common with darker hair synthetic colors and persons with dry absorbent skin. The discoloration will disappear as the skin naturally renews itself and the top layer of skin is sloughed off, typically a few days or at most a week. An easy way to prevent dye discoloration is to put a thin layer of Vaseline or any oil-based preparation around the hairline, and to wear latex or nitrile gloves to protect the hands.

STYLING HAIR

Roberto washes Stephanie's hair to end the coloring process and to begin the styling process. Now there is much discussion between Stephanie and Roberto. Stephanie has been looking through magazines. Roberto is bright-eyed and name drops famous red carpet celebrities. They both agree, definitely not straight, but all curls will look so every day. Stephanie has an idea; she wants her hair swept up and back on one side, to show her diamond studs, while the other side retains a tight curl. A look that will go with the shiny, slinky black dress she is planning to wear to the event tonight. She describes the dress to Roberto and he thinks the look will be brilliant. Stephanie and Roberto confer and decide to blow-dry her hair at the hottest setting using a 3-inch brush, which pulls the wet hair and makes the hair straight as it dries. Once Stephanie's hair is straight, Roberto sweeps the left side of her hair up to the center of her head and braces it with pins. Then begins the process of returning the curls to Stephanie's hair. This will take time. He uses a curling iron to make each lock of hair into a perfectly formed ringlet. Roberto pins some of the ringlets creating cascading curls on the right side of her head and down the nape of her neck. Very glam.

Styling hair really means reshaping the contours of hair. To think about how hair can be styled, let's visualize a thought experiment. A thought experiment is one that you think about, but don't actually do. You will need some imaginary items: a box of spaghetti, water, a black indelible magic marker and a hair dryer. Open the box of spaghetti pasta and take out a fistful. In your fist, notice how each individual spaghetti is straight and lines up in parallel to all the others. Imagine that each spaghetti is a thick strand of hair. When all the spaghettis are straight and parallel to each other, it is like looking at long, straight hair.

With the black magic marker, draw a line across the middle of the group of spaghettis that is perpendicular to the length of the spaghettis. Now, lower the straight pasta into a large container containing room temperature water and try to keep them all parallel. After five minutes, remove the pasta with two hands, place the spaghettis on the counter and bend the entire amount into an "S." Look at the black dots on each spaghetti that were made with the magic marker. They are no longer in a straight line perpendicular to the spaghetti. Each dot on each spaghetti has moved slightly relative to its neighbor. Now, get out the imaginary hair dryer and blow-dry the S-shaped strands of pasta. You have made straight spaghettis into wavy spaghettis. Or, you could have made the wet straight spaghettis into an "e" and heat dried them. That would have given lots of curls of spaghettis. As you can "see," water has allowed you to reshape the spaghettis and heat has set the pasta in its new shape. Pretty simple. It is not much more complicated to style hair. Of course you were just visualizing, but did you notice that when the spaghettis became wet, they swelled and lengthened? Just like hair.

The shape or contour of your hair has a natural default setting: straight, wavy or curly. Styling hair at the hair salon or at home is changing the default setting. As in the spaghetti experiment, water and heat are needed to reshape the natural contours of your hair.

The millions of microfilaments in hair are firmly held together, microfilament to neighboring microfilament, in their default setting by what are called weak chemical bonds, which can be broken by water. Soaking hair in water, washing, shampooing or using a setting lotion, gets water through the close fitting scales of the cuticle and into the cortex. The excess water surrounds the millions of microfilaments inside the hair shaft and the weak chemical bonds break. The wet hair can now be stretched and shaped with pins, clips, rollers, combs, brushes or various hot devices. The next stage of reshaping is to dry the hair into its new style. The hair can be dried straight, wavy or curly. Removing the excess water from the hair by drying permits the weak chemical bonds to form again, but with the microfilaments in a new alignment relative to each other. Applying heat in the form of a hair dryer, hair blower, curling iron or other thermal devices, makes the drying go faster. Alternatively, one can go to sleep with wet hair done up in plastic rollers or toilet paper tubes.

No matter which method used, the weak chemical bonds are being broken and then reset in different places along the microfilaments to give the new look. Getting the hair wet again and then letting it dry without styling, resets the hair back to the natural default setting. The new hairstyle you worked so hard on will wash out during the next shampoo when the weak chemical bonds go back to default due to the water that enters each hair shaft during a shower. The next day, everything must be done again.

Perms and body waves
Sitting next to Stephanie is a teenage girl who hates her naturally straight hair and is getting her first perm. Curly haired women want straight hair; straight haired women want curly hair. The perm, permanent or body wave gives straight haired women a few months of curls and waves.

In 1872, Marcel Grateau offered women the first commercial thermal perm. He devised curved tongs that were heated over a gas flame. He knew the correct temperature was achieved when the tongs turned newspaper brown. Lock by lock, the tongs were applied to hair under tension at successively different angles as the stylist worked their way from near the roots to the tips of the hairs. Essentially, a kind of modern day curling iron without electricity.

When electricity came along, several inventors saw the potential of using electric currents to produce heat in hair curling devices. One of the most successful collaborations was between Eugene Suter and Isidoro Calvete in London. In the 1920's, Eugene, Inc. sold a popular curling machine that looked like an iron octopus hanging from a chandelier with each tentacle holding a section of hair on a woman's head. Twenty-two locks of hair were wound onto curler rods and each curler rod was suspended from above by a wire carrying electricity to produce heat. The curlers were pointed outwards to keep the hot rods away from the scalp for fear of burns. Eventually, many different machines using basically the same principle came into use.

In 1938, Arnold F. Willatt invented the cold wave. This treatment did not need the heat from electricity, but it used chemicals. Alkaline chemicals, as will be described in the next section on relaxers, were used to loosen up the keratin molecules in hair and bend the hair into a new style over a curler. Initially, this process took many hours but, presently, with the use of some heat from a dryer, permanent hair waves can be accomplished with chemicals in less than an hour.

Relaxing hair

Stephanie once tried to relax her hair when she was in her early twenties and wanted long, straight, blonde hair. What she learned was that every morning she needed to spend at least 30 minutes drying and styling her hair. She won't do that again. Too much work for someone in a hurry to get to work in the morning. In any event, she would certainly not get her hair relaxed on the same day it was being colored.

But, there are a lot of women out there, particularly African-American women, who want straight hair. Roberto's hair salon is located just outside Washington, DC and he has a VIP clientele. Sitting across the salon from Stephanie is a woman who looks an awful lot like a high profile Washingtonian. She is having her hair relaxed. Earlier, we discussed hair extensions. Here is the alternative.

A relaxer is a type of lotion or cream that makes hair less curly and easier to straighten by chemically relaxing the natural curls. The active ingredient in these products is lye. Putting lye on people's heads was the brainchild of a creative spirit, Garrett Augustus Morgan, who was born the seventh of eleven children of former slaves. He grew up to be quite an inventor and is probably best known for his invention of the automatic traffic signal and a type of gas mask. Around 1910, he stumbled upon what would become his contribution to the hair care products industry.

While Morgan was working in a sewing machine repair shop, he was trying to come up with a new lubricating liquid for the machine needle. For some reason, he used lye in the lubricating fluid. As the story goes, Morgan wiped the latest version of the goop that was on his hands on a wool cloth and, when he returned the next day, he found the woolly texture of the cloth had smoothed out. He rubbed some goop on an Airedale dog, known for its textured fur, and found he could straighten the wiry, curly fur. Morgan then tried his lubricating liquid on himself. He rubbed it into his hair, felt it burn his scalp and soon washed it out. Lo and behold, his African type hair was not curly anymore; it was straight and looking good. This was his eureka moment and he knew instantly what the goop could be used for. Morgan called it a "hair refining cream" and patented the first chemical hair straightener. He then started a personal grooming products company which included hair dying ointments, curved-tooth pressing combs, shampoo, hair pressing gloss, and the product that started it all, G.A. Morgan's Hair Refiner Cream, which was advertised to "positively straighten hair in 15 minutes."

Lye, also known as sodium hydroxide, is a hazardous caustic alkaline chemical that can dissolve anything from flesh to rock. Lye can burn through paint, metal, cloth, plastic, and especially skin, causing chemical burns, permanent injury or scarring, and blindness. Lye may be fatal if swallowed. In industry, lye is used to make soap, detergents, paper pulp, textiles, drain cleaner and in food processing. To be used as a relaxer, the strong alkali is mixed with petrolatum jelly, mineral oil and emulsifiers to create a creamy consistency. The relaxer is applied to the outer parts of the hair and usually spread around manually by the hairdresser's fingers. Then the relaxer is worked into the roots of the hair and remains in place for a cooking interval or until the scalp burn becomes intolerable.

On application, the caustic cream dissolves away most of the cuticle and permeates into the protein structure of the hair shaft. The disulfide bridges between neighboring keratin molecules are broken by the highly reactive sodium hydroxide and there is probably some actual damage to the proteins themselves. Breaking the microscopic structure of the protofibrils causes the natural curls to loosen up as all the microfilaments in the hair shaft become limp. The hair can be significantly weakened by repeated applications or by a single excessive application, leading to brittleness, breakage, or even hair loss.

Disulfide bridges are hard to break. They are the strong arms sticking out from each keratin molecule and clasping hands with all the other neighboring keratin molecules. Without disulfide bridges, hair would fall apart. Actually, there would be no hair. Even Mother Nature has difficulty breaking disulfide bridges. In fact, there are no natural products or green chemicals that can break disulfide bridges.

At the end of the relaxer treatment, as part of the shampooing, a neutralizer or stabilizer is added to halt the relaxing process and restore the pH balance. This particular step is especially crucial. If not neutralized, the relaxer will continue to work on the hair strands weakening them even further.

From the time the relaxing chemicals are applied to this final neutralizing step, the condition of the hair shafts are extremely fragile and the hair must be handled very carefully. Pulling, tugging, and excessive combing of the hair must be avoided during this period. Afterwards, the prompt use of a hair conditioner is important to replace some of the natural oils that were stripped away by the chemicals in the relaxing process.

Due to increasing public awareness concerning the potential dangers of sodium hydroxide found in traditional relaxer formulas, some women have abandoned them. "No-lye" relaxers have become an alternative

HAIR

for those unwilling to give up relaxers completely. No-lye relaxers do not contain sodium hydroxide but they all use the same mechanism, permeate the hair and break up the disulfide bridges between keratins. No-lye relaxers contain slightly gentler chemicals, such as: potassium hydroxide, lithium hydroxide, guanidine hydroxide or ammonium thioglycolate. The label on these products may be misleading to consumers and try to imply that there aren't any strong chemicals used, or that the chemicals used are somehow less potentially damaging. All of these chemicals are hazardous and not good for hair. In most relaxers sold for home use, the active agent is ammonium sulfite, which also breaks disulfide bridges, but is much weaker and works more slowly. The milder action minimizes, but does not entirely eliminate, collateral irritation to the skin.

Treating hair with keratin

At the same semiformal event that Stephanie was attending, all of the women aged 40-70 looked great. They had long, lustrous hair that they wore straight giving them a youthful appearance. Colors were varied from brownish blonde to auburn reds to soft brunettes. When I introduced myself and the conversation turned to writing a book on hair, big knowing smiles of satisfaction appeared as the women urged me to look into Brazilian keratin hair straightening.

Many of today's high-end popular straighteners are keratin-based. As you have read now many times, keratin is the natural protein found in hair. But, all keratins are not exactly the same. The keratin used in Brazilian keratin hair straightening is extracted from the wool of sheep. Brazilian sheep? Not necessarily but the treatment started in Brazil. The process is supposed to add keratin to hair and it is claimed that the sheep keratin is bonded to human keratin inside the cortex of the hair shaft.

To be chemically reactive, the sheep keratin is mixed with various levels of formaldehyde by the manufacturer. Presumably, formaldehyde helps hold the sheep keratin and the already present keratin molecules together giving the desired straightening. It is also claimed that these products fill in gaps in the hair cuticle that are cracked, dry or damaged.

Concerned about the formaldehyde? You should be. Think corpses and anatomy room cadavers. Formaldehyde fixes tissue so that it does not change further and can be examined by eager first year medical students. A suspected carcinogen, formaldehyde is a colorless chemical compound with a pungent and irritating odor. Exposure to formaldehyde has been linked to health problems, including general malaise, runny nose, sore throat, headache, itching and irritated eyes. The FDA has received myriads of complaints about these keratin products. In the hair industry, concern regarding the use of formaldehyde has been acknowledged and some keratin products are advertised as being formaldehyde-free.

At the hair salon, keratin treatments take about 90 minutes or longer, based on the length of hair, and the results last about two months. The price rises and falls with the hair's length, starting around $300. Color, whether highlights, low lights, or merely covering up the gray, can be done on hair that has had keratin based straightening treatments. Some hair care experts recommend getting the keratin treatment right after coloring in order to seal the dye chemicals in as well.

Are you beginning to see a theme of introducing hair care products containing known hazardous, toxic ingredients that later, after widespread use is established, raises public, governmental and industrial alarm? Be concerned about what you put in your hair.

Stephanie went to the hair salon to overcome her biological heritage and to express her free will. Whether at the hair salon, at the barbershop or at home, we do whatever we want to our hair. Each and every one of us displays a crowning expression of ourselves, and it is uniquely personal and uniquely human.

HAIR

Stephanie's hair, indeed everybody's hair hair, is an unexplained marvel of the human body. Why do we have hair? Hair does not perform vital functions like the heart, the kidneys and the brain. Hair does not help us move around like muscles and bones. Hair does not help us digest food like the stomach and the intestines. What is hair for? We spend a lot of body energy to work those hair follicles steadily for years to grow hair that gets longer and longer. So, why did the furry heads of our ancestor primates become the hairy heads of humans eons ago?

There are a lot of anomalies about hair. By some quirk of evolution, women did not evolve to have hair on their faces but men did. If you shave all of your hair off or go bald, nothing much changes in your life or, at least, in your health. And, what about those European ancestors? Why did their hair become so varied in color and shape? For that matter, why did Asian ancestors develop straight black hair? At least graying is understandable. The hair follicles exhaust their abilities to put color in hair but, in today's world, that occurs rather early in life. Why do men go bald and what's with the characteristic horseshoe shaped pattern of the remaining hair on a bald man's head?

There is no denying that hair is very significant to us. We developed a multi-billion dollar hair care industry to provide us with the resources to take care of our hair. We have not developed a heart care industry or a brain care industry. Something about that visible signal on the top of our head is very important to us. Is it all about physical attraction and sexual selection? Maybe that is what hair is for. Anyway, Stephanie is happy with her uniquely human feature. After that morning at the hair salon, her hair looked MARVELOUS! And it did impress Mr. Right.

My Conclusions

Hair is all about getting the attention of a potential mate, in order to have sex and to make babies. Having done the research for this book, considered all of the available and imaginative material that I could find, and thought about it for several years, my conclusions are that hair developed because of sexual attraction and sexual selection. As Charles Darwin pointed out, these can be powerful forces in human evolution.

Several million years ago, our early ancestors were covered with thick fur, head to foot. They hunted and lived on the African savannas where vast herds of hoofed food roamed. As these hominids evolved as hunters, their legs became stronger, they ran faster, and they used their arms to throw rocks and spears to bring down the game. Whew, did they get hot! So these two-legged hunters lost most of the fur on their heads and bodies, and developed sweat glands. Because their light-colored, relatively furless skin was exposed to the sun, it evolved to be naturally dark to protect them from the rays of the sun. That's how our early ancestors were, up until about one million years ago, dark-skinned with sparse fur, head to foot.

Less than a million years ago, a baby was born in a tribe somewhere on the plains of Africa with a slight mutation of a gene that was to become important in human evolution. The original gene limited the growth of the sparse remaining fur on the head and body to a very short length. The variant gene gave different instructions that lengthened markedly the amount of time hair would grow and, consequently, how long it would grow. And somehow, the variant gene only increased the growing time of hair on the head.

HAIR

For the first year or so, there was nothing unusual about the baby. But as the baby grew, the hair on its head grew longer and longer. This child was different; all the other clan members had short fur on their heads. The tribe members grunted about what to do with this newcomer with long growing hair. Was the child a menacing demon or a harmless oddity? Whatever, the child with the ever-growing hair was not harming anyone, so they let it live. By the time the kid grew up, he or she appeared exotic, different, unusual and sort of attractive because of their long hair. When the leader of the tribe came along and said: "This one is for me," sexual selection started. The evolutionary bandwagon had begun.

For the next hundreds of thousands of years, African type hair was manageable because it generally did not grow very long. But certainly, there were some that found it too long and the clan member who started using sharpened stones or shells to cut it back was appointed tribal barber. Others dressed their hair with wildebeest fat and tied on bones, and hairstyling became an obligatory custom in rituals, ceremonies and dating. Less than one hundred thousand years ago, soon after our ancestors started covering their bodies with clothes, the display on top of their head became the most important, visible attribute of sexual attractiveness. By the time hair evolved into Asian type and European type hair, perhaps thirty to forty thousand years ago, our ancestors were well on their way to caring what their hair looked like.

Hair is our species' sexual display of attractiveness. Hair framing a face, beautiful, handsome or otherwise, is obviously the initial exhibit. We see the hair before we see the eyes or the mouth. Sitting up on top of our head, visible to everyone, we can display our hair in an infinite variety of styles, all of them unique for each individual. We call attention to ourselves by the crowning glory on our heads.

Humans have evolved complex psychological minds so that over time the display of hair has taken on a variety of fashions, styles, conventions and meanings. But, I think that it is safe to say that fur turned into hair because continuously growing, long hair looked good, different, desirable and sexy. For the last ten thousand years, we have been flaunting it.

Afterthoughts

Not too long ago, I was a professor and on the medical school faculty of several top universities. I had large research laboratories and we studied causes and treatments of important eye diseases, like glaucoma and age-related macular degeneration. We published our research findings in scientific journals and took pride in making a contribution to understanding human eye diseases.

If I had it to do all over again, I might work on HAIR. What a fascinating subject. I would set up a research laboratory designed to answer many of the questions raised in this book. I would need a bunch of highly specialized co-workers and technical assistants. These would include an anthropologist, a dermatologist, a molecular biologist, a geneticist, an epidemiologist, an organic chemist, a biochemist, an expert in tissue culture and a physiologist. We would have years of work and require ample funding to take on our subject. Would we get funded? Probably not from government agencies, why would the National Institutes of Health care about hair? But clever and persistent approaches to the companies that make up the hair care industry might get us funds to do our work. They should care about hair.

It is exciting to think about what we could do if we had all the right people together and all the money we needed.

- We could find the genes that regulate the timing of hair growth and figure out how they make hair follicles work for years and years.

HAIR

- We could estimate how long ago in human evolution did the change from fur to hair occur.
- We could grow hair follicles from Africans, Asians and Europeans in tissue culture and find out why the hair shafts come out looking different.
- Perhaps with some genetic manipulations, we could make an African type hair follicle into an Asian type hair follicle that grows an Asian type hair.
- We could determine the functions of the genes in cells in the hair follicle that produce curly, wavy or straight hair.
- We could determine the genetic programs for hair color.
- We could figure out why blonde hair often darkens with age.
- We could find the basis for why there is such variation in the natural color and contour of the hair of Europeans, and why this variation does not occur in Africans or Asians.
- We could determine if the change in the mutated gene, that causes a person to have red hair, affects other tissues in the body.
- To explain graying, we could culture stem cells destined to be melanocytes and figure out why they die off in a relatively young person.
- To possibly explain baldness, we could examine young men with the variant gene for the androgen receptor to determine if they have hyper-androgenetic characteristics when they go through puberty.
- We might be able to find gentler ways to break the disulfide bridges between keratin molecules so that hair relaxes without being damaged.
- I do not know how we would determine the function of the erector pili muscle, but we would think of something.

Of course, all of this is academic research. Probably nothing from our research would help on a bad hair day. But success in this research will tell us more about being human. Maybe I will reopen my laboratory.

APPENDICES

Appendix 1. The Truth about Hair Myths

Shaving off all of a person's hair will make it grow back thicker.
When the hair begins to grow back, it is short and stubby. This only appears to be thicker hair because it is so short. It is not possible to make new hair follicles and shaving the head does not change the structural features of the hair follicles. Shaving a person's head will not make their hair grow back thicker.

Hair continues to grow after death.
At best, the hair might grow $1/50^{th}$ of an inch on a freshly deceased person. This will not be noticeable at the funeral.

Pull out one gray hair and two will grow in its place.
One hair follicle produces one hair shaft at a time. A very simple rule.

Stress produces gray hairs.
There is no medical or scientific evidence that stress causes hair to gray. Nevertheless, we have all seen that the heads of US presidents turn gray over one or two terms of office. This may be due to stress, or it may be due to the natural aging process. A US president usually enters the White House 45-60 years old. During that age range, most people progressively gray whether they are the elected executive or not.

Stress can make your hair fall out.
Again, no compelling medical or scientific evidence. Probably this is self-inflicted.

Severe stress can cause a person's hair to turn white overnight.
As above, there is no medical or scientific evidence that suddenly white can occur. Hair loses its pigment at the base of the hair shaft, not all along the hair shaft. Turning white overnight is not possible.

Redheads have fiery personalities and are sexually wild.
Well, maybe.

As required by their religion, women cover their hair.
The traditional veil used by many Muslim women to cover their hair is called a *hijab*. The use of the *hijab* developed over time in different Muslim cultures and is not universal amongst all Muslim women. The covering of a woman's hair is not mentioned in the Koran. *Hijab* actually means modesty with respect to the Prophet Muhammed's dress code instructions to all of his followers. The Muslim woman is instructed to dress modestly in public and, to many, that includes covering their hair.

Covering hair is not unique to Muslim women. There is some historic evidence that women covered their hair in ancient Greece. In medieval Europe, women of high and low stature covered their hair with veils that were later replaced by hoods. Some orders of Catholic nuns covered their hair as part of their daily dress. Under the nun's headdress, hair was traditionally very short or shorn. Today, these rules have been relaxed. Nevertheless, in many religious groups, women covering their hair while praying is required or encouraged.

Admittedly, a study of stress and graying would be difficult sociological or psychological research to do. Everyone's stress is different.

Appendix 1. The Truth about Hair Myths

Bald men have lots of testosterone and are very virile.
Bald men certainly do not have high levels of circulating testosterone in their blood and, anyway, the amount in the blood is not relevant. Baldness comes on with age; perhaps the virility was at a younger age.

If you are a man and your mother's father was bald, you will be bald.
There is a 50:50 chance that you will be bald, but not a certainty.

Werewolves prowled the streets of London
Hypertrichosis has been suggested as the malady in werewolves in Europe, but no one is certain that werewolves ever existed.

Native Americans cannot grow beards
The generalized belief that Native Americans cannot grow facial hair is a misconception and a stereotype. Although Native Americans are always portrayed without facial hair, most Native Americans can grow a little facial hair above the upper lip and on the chin. Admittedly not much. Some tribes have possessed genetic traits to grow beards, for example men of the Pacific tribes of the western Canadian coast and eastern Alaska have significant facial hair.

Africans cannot grow long hair
African type hair is brittle and breaks easily so that it usually does not grow long. But it can grow long, particularly on admirers of Bob Marley.

Masturbation makes your hair fall out.
Now really!

Appendix 2. Where hair is not wanted:

- In your mouth
- In the drain
- In your food
- On your soap
- On your clothes
- On your pillow
- Out of place
- In your eyes
- On your jacket
- On your lover's jacket
- In your face
- In your computer keyboard
- In your chewing gum
- On the back of your sweaty neck
- When you kiss
- In your comb
- In your brush
- Caught in your barrette
- On your bathroom floor

Appendix 3. Hair Expressions and Phrases

- Hair of the dog: a small alcoholic drink taken to cure a hangover.
- Hair's breadth: a very small amount or margin.
- In someone's hair: to be annoying.
- Let one's hair down: behave in an uninhibited manner.
- Make someone's hair stand on end: alarm or horrify.
- Not a hair out of place: extremely neat in appearance.
- Not turn a hair: remain unmoved.
- Put hair on one's chest: a strong alcoholic drink.
- Split hairs: small and very fine distinctions.
- Took a haircut: reduction of the original value of an asset.
- Hairline fracture: a very thin break in a bone.
- Hairpin turn: a U-shaped curve in a road.
- Hair-raising: extremely alarming.
- Hair shirt: self sacrificing and austere.
- Hair space: the very thin space between letters or words.
- Hair spring: a flat coiled spring regulating the balance wheel in a watch.
- Hair trigger: liable to change suddenly and often explosively.
- By the short hairs: under one's complete control.
- Tear one's hair out: be greatly upset.
- As fine as a frog's hair: feel excellent and in high spirits.
- Acting as a beard: a person who purposely diverts suspicion or attention from another.

Acknowledgements

First, I must acknowledge the Internet, without which this book would probably never have been written. If there were no Internet, taking on a subject that I virtually knew nothing about would have been nearly impossible. The number of hours that I would have spent in libraries would have turned a fun project into a daunting task. Of course, not everything that one finds on the Internet is always true. I have tried to check multiple sources when using Internet information. Sometimes the second and third sources had exactly the same wording as the first source. So be it.

I have to thank Glenn C. Conroy, PhD, a well-known anthropologist at Washington University in St. Louis, for our discussions about hair versus fur, as well as many other subjects. After dinner and a bottle of fine red wine (zinfandel), I made the comment to Glenn that only people have hair that grows long on their head, all other mammals have fur. Glenn's response was that no one had ever distinguished hair from fur as a uniquely human characteristic. Having just discovered a distinguishing feature of being human, we opened another bottle of wine. Sometime later, we wrote up this observation for publication in a scientific journal. The editors of *Evolutionary Anthropology* certainly appreciated our thoughts and accepted our 2005 comment overnight.

I told my children, Erica, Russ and Jess, that after I retired I was going to write a book about everything a person would want to know about hair. For about a year after I retired, they kept asking me when I would start the book. Thanks kids.

I began writing this book during cold, rainy days in the Andes when I was visited by Muse Panguipulli. If it wasn't for the rain and my muse, this book may still not have been started.

Particular thanks to my daughter, Erica, who read and loved all of the chapters, and actually helped me write Chapter 11. Thanks to friends who read chapters here and there, and liked them or hated them. Either way, they motivated me to go further.

Special thanks to Brian Halley of Grub Street for his early editing, and for showing me how to find my voice and write this kind of book. Previously, I had only written scientific papers for technical journals. The writing of this book genre, which I call single subject, nonfiction for non-technical readers, requires a very different style than writing a scientific article. Like using incomplete sentences, colorful adjectives and adverbs effectively, all of which are rarely used in the scientific writing that I practiced throughout my career. And, a certain lightness, which, I hope, helped you get through the biology.

Anyway, it was fun.

References and FYI

This section contains some of the papers published in scientific journals and a few books that I have relied upon. In this digital age of epublishing, ebooks, websites and blogs, I think it is unlikely that most of the readers of this book will go to the library to obtain further information about HAIR. If you are interested, I have also included websites that I used and that will lead you to a world of information on HAIR. Certainly, www.wikipedia.org is a place to start, as I did. The amount of information available from Wikipedia is astounding.

Chapter 1. FROM FURRY HEADS TO HAIRY HEADS
http://onlinelibrary.wiley.com/doi/10.1002/evan.20011/abstract> Human head hair is not fur. The original article published in *Evolutionary Anthropology* by the author and Dr. Glenn C. Conroy which differentiates hair and fur, and states that hair is uniquely human. This book grew from that seed.

Rook's Textbook of Dermatology, eds Burns, Bresthnach, Cox and Griffiths, Vol 4, 7th Edition (2004), Chapter 63, deBerker, Messenger and Sinclair.

http://www.creationresearch.org/crsq/articles/40/40_4/Bergman.htm> Darwinism, creationism and hair.

http://www.nytimes.com/2003/08/19/science/why-humans-and-their-fur-parted-ways.html> Why humans and their fur parted ways. The parasite theory.

Follicular Unit Transplantation. Rousso and Presti. *Facial Plas Surg* (2008) 24:381-388.

Chapter 2. THE HAIR ITSELF

Hair: Its Structure and Response to Cosmetic Preparations. Dawber. *Cli Dermatol* (1996) 14:105-112.

Human Hair: A unique physiochemical composite. Wolfram. *J Am Acad Dermatol.* (2003) 48:S106-114.

Hair follicle-specific keratins and their diseases. Schweizer et al. *Exp Cell Res* (2007) 313:2010-2020.

Cytomechanics of Hair: Basics of the Mechanical Stability. Popescu and Hocker. *Intl Rev Cell Mole Biol* (2009) 277:137-156.

Trichohyalin Mechanically Strengthens the Hair Follicle. Steinert, Parry and Marekov. *J Biochem* (2003) 278: 41409–41419.

http://www.keraplast.com/the-science-behind-replicine> Science of keratin.

Chapter 3. HAIR GROWTH AND THE CONSTRUCTION ZONE

Growth of the Hair. Chase .... (1954) 34:113-126.

Human hair pigmentation – biological aspects. Tobin. *Intl J Cosmetic Sci,* (2008), 30, 233–257.

The Hair Follicle as a Dynamic Mini-organ. Schneider, Schmidt-Ullrich and Paus. *Current Biology* (2009) 19, R132–R142.

The Biology of Hair Care. Draelos. *Dermatologic Clinics* (2000) 18:651-8.

The hair follicle-a stem cell zoo. Jaks, Kasper and Toftgard. *Exp Cell Res* (2010) 316:1422-1428.

http://www.hair-science.com/_int/_en/index.aspx?tc=ROOT-HAIR-SCIENCE^HOME-HAIR-SCIENCE&cur=HOME-HAIR-SCIENCE&> L'Oreal's science site.

http://www.sciencedirect.com/science/article/pii/S0009898106001227> State of the art in hair analysis for detection of drug and alcohol abuse.

Chapter 4. TYPES OF HAIR: FROM TIGHTLY COILED TO STRAIGHT

Shape Variability and Classification of Human Hair: A Worldwide Approach. LaMettrie et al. Human Biology (2007) *Hum Biol* 79:265-281.

Current research on ethnic hair. Franbourg et al. *J Am Acad Dermatol* (2003) S115-119.

Evolutionary aspects of hair polymorphism. Chernova. (2006). *Biol Bull* 33, 43–52.

A scan for genetic determinants of human hair morphology: EDAR is associated with Asian hair thickness. Fujimoto et al. (2008). *Hum Mol Genet* 17, 835–843.

FGFR2 is associated with hair thickness in Asian populations. Fujimoto et al. (2009) *J Hum Genet* 54, 461–465.

Common Variants in the Trichohyalin Gene Are Associated with Straight Hair in Europeans. Medland et al. *Am J Hum Gen* (2009) 85, 750–755.

Apparent fragility of African hair is unrelated to the cystine-rich protein distribution: a cytochemical electron microscopy study. Khumalo, Dawber and Ferguson. *Exp Dermatol* (2005) 14:311-314.

Hair Care Practices in African American Patients, Roseborough and McMichael. *Semin Cutan Med Surg* (2009) 28:103-108.

Human hair shape is programmed from the bulb. Thibaut et al. *Br J Dermatol* (2005) 152:631-8.

http://www.nytimes.com/2011/05/17/us/17hair.html?emc=eta1> Stealing hair extensions.

"Good Hair," movie/DVD by Chris Rock, 2009.

http://www.macmillandictionary.com/thesaurus-category/british/Words-used-to-describe-the-state-of-people-s-hair> This site contains 56 words describing hair.

Chapter 5. NATURAL HAIR COLOR -OR- WHY ARE THERE REDHEADS?

The physical and chemical properties of eumelanin. Meredith and Sarna. *Pigment Cell Res* (2006) 19:572-594.

Current challenges in understanding melanogenesis. Simon et al, *Pigment Cell Melanoma Res* (2009) 22:563-579.

Human hair pigmentation-biological aspects. Tobin. *J Cosmetic Sci* (2008) 30:233-257.

European hair and eye color - A case of frequency-dependent sexual selection? Frost. *Evolution and Human Behavior* (2006) 27:85-103.

Worldwide polymorphism at the MC1R locus and normal pigmentation variation in humans. Makova and Norton. *Peptides* (2005) 26:1901-1908.

Seeing red: pheomelanin synthesis uncovered. Simon. *Pigment Cell Melanoma Res* (2009) 22:382-383.

The Melacortin 1 Receptor (MC1R): More Than Just Red Hair. Rees. *Pigment Cell Melanoma Res* (2000) 13;135-140.

A Draft Sequence of the Neandertal Genome. Green et al. *Science* (2010) 328, 710.

Model-based prediction of human hair color using DNA variants, Branicki et al. *Hum Genet* (2011) 129:443–454.

The Roots of Desire: the myth, meaning and sexual power of red hair. Marion Roach, 2006, Bloomsbury USA.

Still Life with Woodpecker. Tom Robbins, 1980, Bantam.

Chapter 6. THE GRAYING OF THE HEADS OF THE WORLD

Graying: gerontobiology of the hair follicle pigmentary unit. Tobin and Paus. *Exp Gerontology* (2001) 36:29-54.

Aging, Graying and Loss of Melanocyte Stem Cells. Sarin and Artandi. *Stem Cell Res* (2007) 3:212-217.

Melanocyte Stem Cell Maintenance and Hair Graying. Steingrmisson, Copeland and Jenkins. *Cell* (2005) 121:9-12.

Rate of Greying of Human Hair. Keough and Walsh. *Nature* (1965) 207:877-878.

Towards a "free radical theory of graying. *The FASEB J* (2006) 20:1567-1569.

Pharmacologic Interventions in aging hair. Trueb. *Clin Interventions in Aging* (2006) 1:121-129.

Marie Antoinette Syndrome. Navarini, Nobbe and Trueb. *Arch Dermatol* (2009) 145:656.

Chapter 7. BALDING (HIS AND HERS)

The effects of hair loss in European men: a survey in four countries. Budd et al. *Europ J Dermatol* (2000) 10:122-127.

Does fortune favour the bald? Psychological correlates of hair loss in males. Wells, Willmoth and Russell. *Br J Psychol* (1995) 86:337-344.

Genetic Variation in the Human Androgen Receptor Gene Is the Major Determinant of Common Early-Onset Androgenetic Alopecia. Hillmer et al. *Am J Hum Genetics* (2005) 77:140-148.

Genome-wide Scan and Fine-Mapping Linkage Study of Androgenetic Alopecia Reveals a Locus on Chromosome 3q26. Hillmer et al. *Am J Human Genetics* (2008) 82:737-743.

Androgens and hair loss. Alsantali and Shapiro. *Curr Opinion Endocinol Diab Obes* (2009) 16:246-253.

EDA2R Is Associated with Androgenetic Alopecia, Prodi et al. *J Invest Derm* (2008) 128, 2268–2270.

Baldness may be caused by the weight of the scalp: Gravity as a proposed mechanism of hair loss. Ustuner. *Med Hypotheses* (2008) 71:505-514.

http://well.blogs.nytimes.com/2010/01/15/when-women-lose-their-hair/?emc=eta1> Hair loss in women.

http://www.hairclub.com/> Treatments for male and female hair loss, mostly surgical.

http://well.blogs.nytimes.com/2011/02/16/in-surprise-finding-bald-mice-find-their-fur-again/?emc=eta1> Bald mice grow hair. Is there hope for a new solution?

http://www.moviestar.nu/articles/baldmoviestarsenglish.html> Secretly balding movie stars.

http://www.thebaldtruth.com/> All sorts of information about balding and what can be done about it, comments, complaints and a radio show.

Chapter 8. FACIAL HAIR (HIS AND HERS)

http://www.ftmguide.org/facialhair.html> Growing and grooming a beard.

http://artofmanliness.com/2008/11/02/20-manliest-mustaches-and-beards-from-facial-hair-history/> Manliness and moustaches and beards.

http://www.britishpathe.com/record.php?id=26183> A video from 1923 of Hans Langseth, the man with the longest beard ever, playing with his beard.

Different growth rates of pigmented and white hair in the beard. Nagl. *Br J Dermatol* (1995) 132:94-7.

www.beard.org> Everything you need to know about growing, coloring and grooming your beard.

Chapter 9. HAIR CARE, TRICHOLOGY, FUNGI AND NITS

Hair and Scalp Diseases in *Basic and Clinical Dermatology*, eds McMichael and Hordinsky, 2008, Informa healthcare.

Shampoos: Ingredients, efficacy and adverse effects. Trueb. *JDDG* (2007) 5:356-365.

http://www.thedailybeast.com/newsweek/blogs/the-human-condition/2009/10/08/the-science-of-shampoo-what-the-ingredients-mean.html> The science of shampoos.

http://www.independent.co.uk/news/science/the-truth-about-the-science-bit-733265.html> Shampoo and hair care companies.

http://sciencemags.blogspot.com/2010/01/hairs-shampoo-and-conditioner.html> Conditioners.

http://www.trichology.edu.au/> The International Association of Trichologists.
http://www.webmd.com/skin-problems-and-treatments/hair-loss/infectious-agents> Fungi in hair.

Chapter 10. CUTTING HAIR AND BLOODLETTING
http://wordinfo.info/unit/3364/ip:17> barber history.
http://www.pbs.org/wnet/redgold/basics/bloodletting.html> PBS documentary about bloodletting.
http://www.utexas.edu/opa/blogs/culturalcompass/2011/01/13/locks-of-ages-the-leigh-hunt-hair-collection/> The Leigh Hunt hair collection features locks from 21 authors and statesmen, including John Milton, John Keats, and George Washington.
http://www.cosmetologybarberingschools.com> How to start a new profession.

Chapter 11. A MORNING AT THE HAIR SALON
http://science.howstuffworks.com/innovation/everyday-innovations/hair-coloring.htm> Hair coloring.
http://www.nytimes.com/2011/06/09/fashion/hair-care-for-african-americans.html?emc=eta1> African-American women and their quest for natural hair.
http://well.blogs.nytimes.com/2011/08/25/surgeon-general-calls-for-health-over-hair/?emc=eta1> The Surgeon General of the US calling for healthy hair.
"Relaxers" damage hair: Evidence from amino acid analysis. Khumalo et al. *J Am Acad Dermatol* (2010) 62:402-8.
http://luciagabriela.ws/keratin-straightening-science/> Straightening hair with keratin.
http://articles.chicagotribune.com/2011-01-26/health/ct-x-n-keratin-hair-treatment-20110126_1_brazilian-blowout-keratin-complex-hair-smoothing-treatments> Warnings about the use of keratin to straighten hair.

www.ingramcontent.com/pod-product-compliance
Lightning Source LLC
Chambersburg PA
CBHW030934180526
45163CB00002B/564